64

Advances in Biochemical Engineering/Biotechnology

Managing Editor: T. Scheper

Springer-Verlag Berlin Heidelberg GmbH

Thermal Biosensors
Bioactivity
Bioaffinity

With contributions by
P. K. Bhatia, B. Danielsson, P. Gemeiner,
S. Grabley, F. Lammers, A. Mukhopadhyay,
K. Ramanathan, M. Saleemuddin, T. Scheper,
V. Stefuca, R. Thiericke, B. Xie

Springer

Advances in Biochemical Engineering/Biotechnology reviews actual trends in modern biotechnology. Its aim is to cover all aspects of this interdisciplinary technology where knowledge, methods and expertise are required for chemistry, biochemistry, microbiology, genetics, chemical engineering and computer science. Special volumes are dedicated to selected topics in which the interdisciplinary interactions of this technology are reflected. New biotechnological products and new processes for synthesizing and purifying these products are at the center of interest. New discoveries and applications are discussed.

In general, special volumes are edited by well known guest editors. The managing editor and publisher will however always be pleased to receive suggestions and supplementary information. Manuscripts are accepted in English.

In references Advances in Biochemical Engineering/Biotechnology is abbreviated as Adv. Biochem. Engin./Biotechnol. as a journal.

ISSN 0724-6145
ISBN 978-3-662-14727-6 ISBN 978-3-540-49811-7 (eBook)
DOI 10.1007/978-3-540-49811-7

Library of Congress Catalog Card Number 72-152360

Originally published by Springer-Verlag Berlin Heidelberg New York in 1999
Softcover reprint of the hardcover 1st edition 1999

Typesetting: Fotosatz-Service Köhler GmbH, Würzburg
Cover: Design & Production, Heidelberg
SPIN: 10648321 02/3020 - 5 4 3 2 1 0 – Printed on acid-free paper

Attention all "Enzyme Handbook" Users:

Information on this handbook can be found via the internet at

http://www.springer.de/chem/samsup/enzym-hb/ehb_home.html

At no charge you can download the complete volume indexes Volumes 1 through 13 from the Springer www server at the above mentioned URL. Just click on the volume you are interested in and receive the list of enzymes according to their EC-numbers.

Contents

Principles of Enzyme Thermistor Systems: Applications to Biomedical and Other Measurements

Bin Xie · Kumaran Ramanathan · Bengt Danielsson

Department of Pure and Applied Biochemistry, Lund University, Box 124, Center for Chemistry and Chemical Engineering, S-221 00 Lund, Sweden. *E-mail: Bengt.Danielsson@tbiokem.lth.se*

This chapter presents an overview of thermistor-based calorimetric measurements. Bioanalytical applications are emphasized from both the chemical and biomedical points of view. The introductory section elucidates the principles involved in the thermometric measurements. The following section describes in detail the evolution of the various versions of enzyme-thermistor devices. Special emphasis is laid on the description of modern "mini" and "miniaturized" versions of enzyme thermistors. Hybrid devices are also introduced in this section. In the sections on applications, the clinical/biomedical areas are dealt with separately, followed by other applications. Mention is also made of miscellaneous applications. A special section is devoted to future developments, wherein novel concepts of telemedicine and home diagnostics are highlighted. The role of communication and information technology in telemedicine is also mentioned. In the concluding sections, an attempt is made to incorporate the most recent references on specific topics based on enzyme-thermistor systems.

Keywords: Enzyme thermistor, Calorimetry, Clinical, Telemedicine, Home diagnostics.

Advances in Biochemical Engineering/
Biotechnology, Vol. 64
Managing Editor: Th. Scheper

1 Introduction

1.1 Fundamentals of Calorimetric Devices

Life is made up of lifeless molecules, but when the molecules react with each other there is an exchange of several forms of energy. One well-known form of energy is heat. The merits of measuring heat (calorimetry) were identified several decades ago. Almost all effects, either physical, chemical or biological involve exchange of heat. Specifically, biological reactions involving enzyme catalysis are associated with rather large enthalpy changes. Based on the nature of the catalytic reaction, either a single enzyme or a combination of enzymes can be employed for generating a detectable thermal signal.

In earlier investigations a wider application of calorimetry, especially in routine analysis, was limited due to the need for sophisticated instrumentation, the relatively slow response and the high costs [1]. Several simple calorimetric devices based on immobilized enzymes were introduced in the early 1970s that combined the general detection principle of calorimetry with enzyme catalysis [2]. The advantages of these instruments were reusability of the biocatalyst, the possibility of continuous flow operation, inertness to optical and electrochemical interference, and simple operating procedures. Several of these concepts, developed in the following years, culminated in the development of the enzyme thermistor (ET) designed in our laboratory. The technique drew immediate attention in the area of biosensors and has been successfully exploited in the last two decades. The initial biosensing applications focused on the determination of glucose and urea, and has subsequently been applied in the determination of a wide variety of molecules [3].

Development of simple, low-cost calorimeters for routine analysis, called thermal enzyme probes (TEP), has been attempted by several groups. These are fabricated by attaching the enzyme directly to a thermistor [4, 5]. However, in this configuration, most of the heat evolved in the enzymic reaction is lost to the surrounding, resulting in lower sensitivity. The concept of TEP was essentially designed for batch operation, in which the enzyme is attached to a thin aluminum foil placed on the surface of the Peltier element that acts as a temperature sensor [6].

Although in later designs [7, 8], the sensitivity of TEP was improved, considerable enhancement in detection efficiency was achieved by employing a small column, with the enzyme immobilized on a suitable support. In this case, the heat is transported by the circulating liquid passing through the column to a temperature sensor mounted at the top of the column. Several models of this configuration were developed in the mid-1970s, including the *enzyme thermistor* and the *immobilized enzyme flow-enthalpimetric analyzer* [9, 10]. Furthermore, a commercial flow-enthalpimeter combined with an immobilized enzyme column has also been described [11].

Recently, several miniaturized prototypes have been fabricated, e.g., a thermal probe for glucose, designed as an integrated circuit, called a biocalorimetric sensor, with total dimensions of only 1 × 1 × 0.3 mm [12]. In a different model, a small thermoelectric glucose sensor employing a thin-film thermopile to measure the evolved heat was described. These devices were reported to be less affected by external thermal effects compared to thermistor based calorimetric sensors and could be operated without environmental temperature control [13]. Active work is in progress in the authors' laboratory to construct a miniaturized portable biothermal flow-injection system suitable for on-line monitoring. An instrument with 0.1 – 0.2 mm (ID) flow channels and a flow rate of 25 – 30 μl/min with sample volumes of 1 - 10 μl is being evaluated at present. A 1 × 5 mm enzyme column allows determination of glucose concentrations down to 0.1 mM. Recently a device equipped with thin-film temperature sensors of thermistor type (0.1 × 0.1 mm or smaller) for glucose measurements has also been developed [14].

1.2 Principle of Calorimetric Measurement

The total heat evolution is proportional to the molar enthalpy and to the total number of product molecules created in the reaction.

$$Q = -n_p(\Delta H)$$

where Q = total heat, n_p = moles of product, and ΔH = molar enthalpy change. It is also dependent on the heat capacity C_p of the system, including the solvent:

$$Q = C_p(\Delta T)$$

The change in temperature (ΔT) recorded by the ET is thus directly proportional to the enthalpy change and inversely proportional to the heat capacity of the reaction.

$$\Delta T = -\Delta H \, n_p / C_p$$

As the heat capacity of most organic solvents is two or three times lower than that of water, enhanced sensitivity is expected in organic solvents. This is the case, provided that ΔH remains unaltered. This area of investigation is dealt with in detail in a later section.

Table 1 presents a list of the molar enthalpy changes in a few enzyme-catalyzed reactions. A thermometric measurement is based on the sum of all enthalpy changes in the reaction mixture. Thus, it is favorable to co-immobilize oxidases with catalase, which results in doubling the sensitivity, nullifying the deleterious effects of hydrogen peroxide, and simultaneously reducing the oxygen consumption. As indicated in Table 1, the high protonation enthalpy of buffer ions like Tris can be utilized to enhance the total enthalpy of proton-producing reactions. A notable increase in the sensitivity can also be obtained in substrate- or coenzyme-recycling enzyme systems, in which the net enthalpy change of each

Table 1. Linear concentration ranges of substances measured with calorimetric sensors using immobilized enzymes

Analyte	Enzyme(s) used	Linear range (mM)	Enthalpy change (–kJ/mol)
Ascorbic acid	Ascorbate oxidase	0.01–0.6	
ATP (or ADP)	Pyruvate kinase + hexokinase	10 nM[a]	
Cellobiose	β-Glucosidase + glucose oxidase/catalase	0.05–5	
Cholesterol	Cholesterol oxidase/catalase	0.01–3	53 + 100
Creatinine	Creatinine iminohydrolase	0.01–10	
Ethanol	Alcohol oxidase	0.0005–1	
Glucose	Hexokinase	0.01–25	28 (75[b])
Glucose	Glucose oxidase/catalase	0.0002–1 –75[c]	80 + 100
L-Lactate	Lactate-2-mono-oxygenase	0.005–2	
L-Lactate	Lactate oxidase/catalase	0.0002–1	
L-Lactate (or pyruvate)	Lactate oxidase/catalase + lactate dehydrogenase	10 nM[a]	ca. 225
Oxalate	Oxalate oxidase	0.005–0.5	
Penicillin G	β-Lactamase	0.002–200	67 (115[b])
Pyruvate	Lactate dehydrogenase	0.01–10	47 (15[b])
Sucrose	Invertase	0.05–100	
Urea	Urease	0.005–200	61

[a] With substrate recycling. [b] In Tris buffer, [c] With benzoquinone as electron acceptor.

turn in the cycle adds to the overall enthalpy change [15]. A later section provides further details on chemical and enzymatic amplification.

An inherent disadvantage of calorimetry is the lack of specificity: All enthalpy changes in the reaction mixture contribute to the final measurement. It is therefore essential to avoid nonspecific enthalpy changes due to dilution or solvation effects. In most cases this is not a serious problem. An efficient way of coping with nonspecific effects in a differential determination is the incorporation of a reference column with an inactive filling [16].

The flow injection technique is usually employed for an ET assay. The sample volumes employed are too small to produce a thermal steady state, but generate a temperature peak. This is traced by a recorder. The peak height of the thermometric recording is proportional to the enthalpy change corresponding to a specific substrate concentration. In most instances, the area under the peak and the ascending slope of the peak have also been found to vary linearly with the substrate concentration [17]. A sample introduction of sufficient duration (several minutes) leads to a thermal steady state resulting in a temperature change, proportional to the enthalpy change up to a certain substrate concentration [62].

1.3 The Transducer

The instrumentation for fabrication of the ET normally employs a thermistor as a temperature transducer. Thermistors are resistors with a very high negative temperature coefficient of resistance. These resistors are ceramic semiconductors, made by sintering mixtures of metal oxides from manganese, nickel, cobalt, copper, iron and uranium. They can be obtained from the manufacturers in many different configurations, sizes (down to 0.1–0.3 mm beads) and with varying resistance values The best empirical expression to date describing the resistance-temperature relationship is the Steinhart-Hart equation:

$$1/T = A + B\,(\ln R) + C\,(\ln R)^3$$

where T = temperature (K); ln R = the natural logarithm of the resitance, and A, B and C are derived coefficients. For narrow temperature ranges the above relationship can be approximated by the equation:

$$R_T = R_{To}\, e^{\beta(1/T - 1/To)}$$

where R_T and R_{To} are the zero-power resistances at the absolute temperatures T and T_0, respectively, and β is a material constant that ranges between 4000 and 5000 K for most thermistor materials. This yields a temperature coefficient of resistance between -3 and -5.7% per °C. In our ET devices resistances of 2–100 kΩ have been used. Other temperature transducers employed in enzyme calorimetric analyzers include Peltier elements, Darlington transistors, and thermopiles. Of these, the thermistor is the most sensitive of the common temperature transducers.

2 Description of Instrumentation and Procedures

2.1 Conventional System

The earlier investigations employed several different types of plexiglas constructions containing the immobilized enzyme column. These devices were thermostated in a water bath, and the temperature at the point of exit from the column was monitored with a thermistor connected to a commercial Wheatstone bridge. The latter was constructed for general temperature measurements and osmometry. Later, we developed more sensitive instruments for temperature monitoring indigenously: the water bath was replaced by a carefully temperature-controlled metal block, which contained the enzyme column. The enzyme thermistor concept has been patented in several major countries.

Such simple plexiglas devices were extremely useful and could be employed for determinations down to 0.01 mM. An example of such a simple device will therefore be described here in some detail. (Fig. 1). The plastic column, which can hold up to 1 ml of the immobilized enzyme preparation, was mounted into a plexiglas holder, leaving an insulating airspace around the column. The heat exchanger consisted of acid-proof steel tubing (ID 0.8–1 mm and about 50 cm long) coiled and placed in a water-filled cup. The whole device was placed in a water bath with a temperature stability of at least 0.01 °C. The cap surrounding the heat exchanger reduced the temperature fluctuations considerably and improved the baseline.

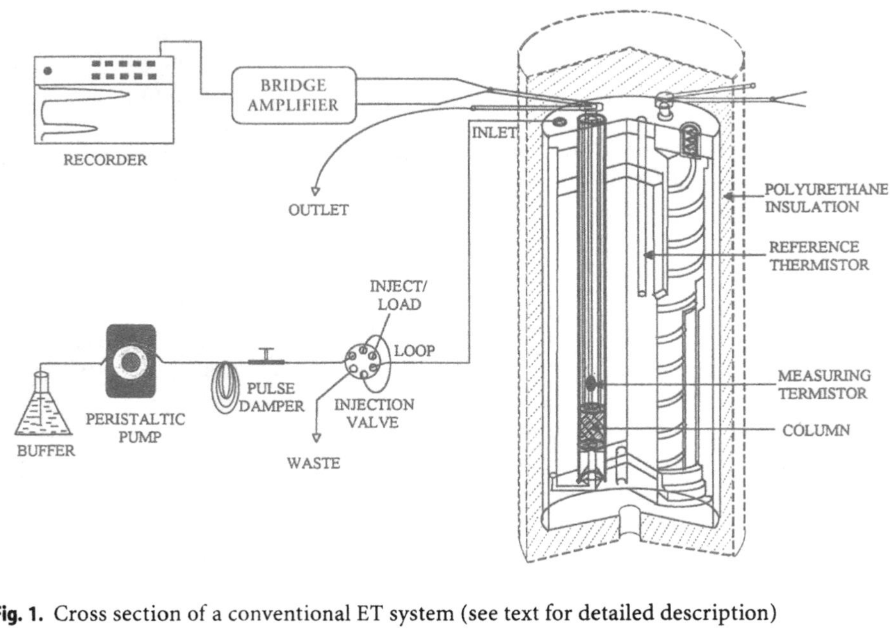

Fig. 1. Cross section of a conventional ET system (see text for detailed description)

The temperature was measured at the top of the column with a thermistor (10 kΩ at 25 °C, 1.5 × 6 mm, or equivalent) epoxied at the tip of a 2 mm (OD) acid-proof steel tube. The temperature was measured as an unbalance signal of a sensitive Wheatstone bridge. At the most sensitive setting, the recorder output produced 100 mV at a temperature change of 0.01 °C. Placing the temperature probe at the very top of the column, rather than in the effluent outside the column, reduced the turbulence around the thermistor and gave a more stable temperature recording.

The solution was pumped through the system at a flow rate of 1 ml/min with a peristaltic pump. The sample (0.1 – 1 ml) was introduced with a three-way valve or a chromatographic sample loop valve. The height of the resulting temperature peak was used as a measure of the substrate concentration and was found to be linear with substrate concentration over a wide range. Typically it was 0.01 – 100 mM, if not limited by the amount of enzyme or deficiency in any of the reactants.

For example, this type of instrument was adequate for the determination of urea in clinical samples. The sensitivity was high enough to permit a 10-fold dilution of the samples, which eliminates problems of nonspecific heat. The resolution was consequently about 0.1 mM, and up to 30 samples could be measured per hour.

In order to achieve more sensitive determinations, we developed a two-channel instrument, in which the water bath was replaced by a carefully thermostated metal block. A specially designed Wheatstone bridge permits temperature determinations with a sensitivity resolution of 100 mV/0.001 °C. The calorimeter was placed in a container insulated by polyurethane foam. It consists of an outer aluminum cylinder which can be thermostated at 25, 30 or 37 °C with a stability of at least ± 0.01 °C. Inside is a second aluminum cylinder with channels for two columns and a pocket for a reference thermistor. The solution passes through a thin-walled acid-proof steel tube (ID 0.8 mm) before entering the column. Two-thirds of this tubing acts as a coarse heat exchanger in contact with the outer cylinder, while the remaining third is in contact with the inner cylinder. This has a higher heat capacity and is separated from the thermostated jacket by an airspace. Consequently, the column is held at a very constant temperature, and the temperature fluctuations of the solution become exceedingly small.

The columns were attached to the end of the plastic tubes by which they are inserted into the calorimeter. Columns could therefore be readily changed with a minimum disturbance of the temperature equilibrium. Inside the plastic tube were the effluent tubing and the leads to the thermistor were fastened to a short piece of gold capillary with heat-conducting, electrically insulating epoxy resin. Veco Type A 395 thermistors (16 kΩ at 25 °C, temperature coefficient 3.9 %/ °C) were used. These are very small, dual-bead isotherm thermistors with 1 % accuracy; as such, they were interconvertible, comparatively well matched, and follow the same temperature response curve (within 1 %). An identical thermistor was also mounted in the reference probe.

The Wheatstone bridge was built with precision resistors that have a low temperature coefficient (0.1 %; temperature coefficient 3 ppm) and was equipped with a chopper-stabilized operational amplifier. This bridge produced a maximum change of 100 mV in the recorder signal for a temperature change of 0.001 °C. However, the lowest practically useful temperature range was, limited mainly by temperature fluctuations caused by friction and turbulence in the column: typically 0.005 – 0.01 °C. Differential temperature recordings were made against a reference thermistor that was inserted in a pocket in the inner aluminum block or against an identical thermistor probe with an inactive reference column. The latter arrangement was useful when nonspecific heat effects (e.g., due to heat of solvation or dilution) were encountered. The sample was then split equally between the enzyme column and the reference column [18]. Alternatively, the second channel could be reserved for another enzyme preparation, permitting a quick change of enzymatic analysis. Some instruments were even equipped with a dual Wheatstone bridge, enabling two different, independent analyses to be carried out simultaneously. In such conventional systems it is an advantage to replace used-up enzyme columns quickly, although the fresh column requires about 20 – 30 min to achieve thermal equilibrium with the circulating buffer. In total, over 30 instruments of the newer type have been assembled at the work-shop of our centre for use in industry and in research institutes.

A major drawback has been the unwillingness of the public to accept an uncommon technique, resulting in fewer developments. The common mode rejection (detection is non specific), non discriminative, compared to spectroscopic (wherein specific wavelength is used) and electrochemical (wherein specific potential is used) for detection. In addition the technique is not very sensitive. The theoretical sensitivity is 0.1 µM using the GOX/catalase system.

There are other more sensitive techniques, such as amperometry chemiluminescence and fiber fluorimetry, that have recently been developed in our laboratory which are not as robust as the ET. The high concentration of enzyme employed in the column may be treated both as an advantage and disadvantage. This increases the operational stability of the column to over a couple of years, compared to a few months with lower enzyme loads. The transducer (thermistor) has no fouling or drift, thus making the calibration extremely simple. Such an approach is useful for bioprocess monitoring and scaling up or down is simple. The general approach can be used for a multiple enzyme system with thermistors at the inlet and outlet of each enzyme block, and the carry-over heat – if any – can be subtracted for each enzyme/substrate combination. Oxygen (for oxidase reactions) can be regenerated by electrolysis of water, with the use of platinum electrodes, directly in the flow stream in the vicinity of the enzyme. Similar approaches had already been demonstrated by us. The pH changes – if any – are negligible and well within the buffering capacity of the circulating buffer, and do not pose a major problem while using multiple enzymes.

2.2 Minisystem

The design of mini and micro systems calls for a multidisciplinary approach involving engineering, materials science, electronics and chemistry. With the advances in integrated circuit technology and micromachining of liquid filters,

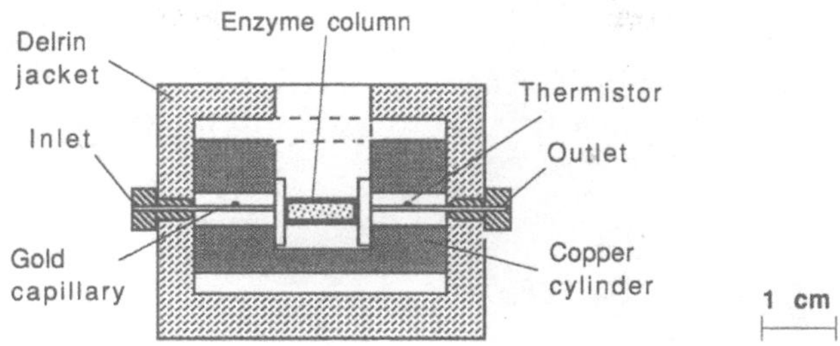

Fig. 2. Schematic cross section of a compact mini ET system (see text for detailed description)

transducers, microvalves and micropumps on chips, practically useful microsystem technology has become a reality [19].

A compact sensor of greatly reduced dimensions (outer diameter × length: 36 × 46 mm) has been constructed and is shown in Fig. 2. In order to conveniently accommodate enzyme columns and to ensure isolation from ambient temperature fluctuations, a cylindrical copper heat sink was included. An outer Delrin jacket further improved the insulation. The enzyme column (inner diameter × length: 3 × 4 mm), constructed of Delrin, was held tightly against the inner terminals of the copper core. Short pieces of well-insulated gold capillaries (outer diameter/inner diameter: 0.3/0.2 mm) were placed next to the enzyme column as temperature-sensitive elements. Microbead thermistors were mounted on the capillaries with a heat-conducting epoxy. Two types of mini system has been constructed as discussed below.

2.2.1
Plastic Chip Sensor

The chip sensor (27 × 7 × 6 mm) was constructed of Plexiglas (see Fig. 3). The rectangular enzyme cell (5 × 3 × 0.5 mm) and the inlet and outlet parts were

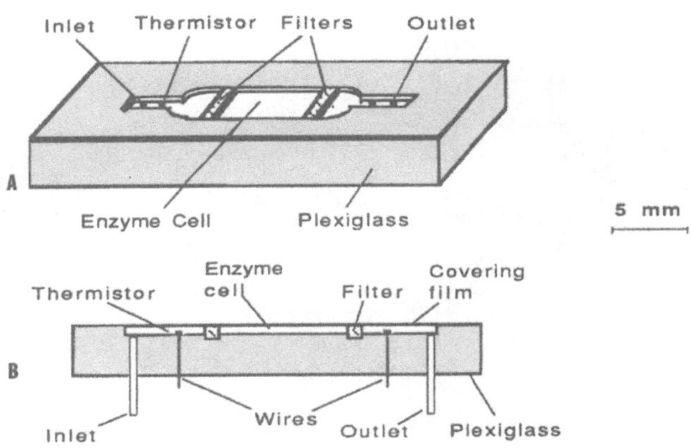

Fig. 3. Schematic diagram of a microcolumn ET sensor (see text for detailed description)

milled into the plastic base to a depth of 0.5 mm. Electrically-insulated thermistors in direct contact with flow stream were placed outside the enzyme cell after the porous polyethylene filters. The enzyme cell was charged with the enzyme preparation prior to compaction. Ready access to the enzyme compartment makes replacement of enzymes possible.

2.2.2
Microcolumn Sensor

The microcolumn sensor (inner diameter x length: 0.6 × 15 mm) was constructed of stainless-steel tubing and is free of auxiliary components (see Fig. 4). Microbead thermistors (thermistors shaped like a bead) were directly mounted in the reference and measurement probes on the outer surface of the inlet and outlet gold tubings, using heat-conducting epoxy. Both the length (about 200 mm) and inner diameter (0.15 mm) of the inlet tubing between the sample valve and the column were minimized, in order to reduce sample dispersion during transport in the flow system. Similarly, in order to reduce heat loss, simple and short connections between the column and the inlet or outlet tubings were required.

2.3
Microsystems

A planar substrate, such as silicon wafer, could be micromachined by a sequence of deposition and etching processes. This results in three-dimensional microstructures which can be implemented in cavities, grooves, holes, diaphragms, cantilever beams etc. The process referred to as silicon micromachining often employs anisotropic etchants such as potassium hydroxide and ethylene diamine pyrocatechol. The crystallographic orientation is important as the above-mentioned etchants show an etch-rate anisotropy. The ratio for the (100)-, (110)- and (111)- planes is typically 100 : 16 : 1. The technique of electrochemical etch stop could be applied for control of the microstructural dimensions. An alternative

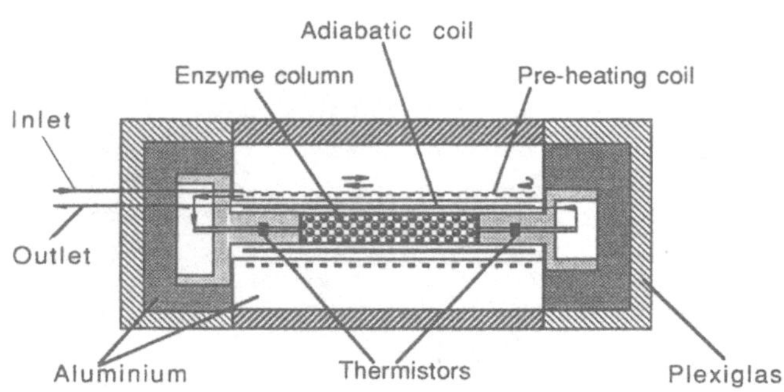

Fig. 4. Illustration of a miniaturized ET system (refer to text for details)

technique is surface micromachining, in which a sacrificial layer is selectively etched from below an etch-resistant thin film. Static and dynamic micromechanical structures have been fabricated by applying these processes in combination with low-pressure chemical vapour deposition on polysilicon. Such systems, i.e. that have the electronic properties of semiconductors, are comparable with thermistors, where there is a change in resistance as a function of temperature, or a thermopile based on the Seebeck effect or the *p*-n junction for the diode and transistor. In addition, integrated thermistors and thermopiles can be designed by doping boron into polysilicon, in order to achieve a temperature dependent change in resistance or to form a thermocouple in the presence of aluminium or gold.

An additional advantage of the integrated circuit technology is the ability to integrate the various components, such as the transducer, reactor, valve, pump etc., within the electronic system, forming refined flow-analysis systems on silicon wafers. Several approaches, such as electrostatic, electromagnetic, piezoelectric, thermopneumatic and thermoelectric can be employed for force transduction in the microvalves, these are also applicable to micropumps. Based on these approaches, two versions of micropumps have been developed. These are connected in parallel; the first pump (dual pump) is activated with periodic two-phase voltage, while the second pump (the buffer pump) is driven by two piezoelectric actuators. Microsensors of two kinds are described below: a thermopile based- and a thermistor based microbiosensor.

2.3.1
Thermopile-Based Microbiosensor

The thermopile-based microbiosensor (Fig. 5) is fabricated on a quartz chip. Its functioning is based on the Seebeck effect: $\Delta V = n\,\alpha_{ab}\,\Delta T$, where ΔV is the voltage output of one thermocouple; n stands for the number of thermocouples, ΔT is the temperature difference between the hot junction and the cold junction, and α_{ab} is the relative Seebeck coefficient, which is dependent on the composition of the material and on the working temperature. For small temperature ranges, the Seebeck coefficient α_{ab} can be considered to be constant. Thus, the voltage output of the thermocouple is proportional to the temperature difference, ΔT between the hot and the cold junctions. A thermopile was constructed by connecting a number of thermocouples in series. The thermopile has a much larger voltage output than a single thermocouple for the same temperature difference, since the output from the thermopile is equal to the sum of the outputs from each thermocouple. When the cold junction is maintained at a constant temperature and the hot junction is placed proximally to the exothermic enzyme reaction, the detection of the output voltage from the thermopile is directly related to the substrate concentration.

The integrated thermopile (1.6×10 mm) was manufactured by the following method. A quartz chip ($25.2 \times 14.8 \times 0.6$ mm) was used as a substrate instead of a silicon wafer, in order to reduce the heat conductivity of the chip. A 0.5 µm thick layer of polysilicon was deposited using LPCVD (low-pressure chemical vapour deposition) onto the quartz substrate. The layer was boron-doped using

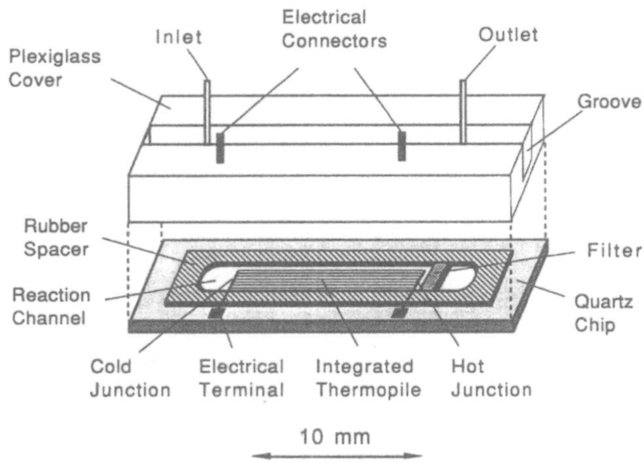

Fig. 5. Schematic diagram of the thermal microbiosensor based on an integrated thermopile fabricated on a quartz chip. Glucose oxidase immobilized on CPG beads was charged into the channel

ion implantation and then annealed in nitrogen at 950 °C for 30 min. Next, the layer was patterned by wet chemical etching, using negative photoresist as an etch mask. Metalization was accomplished through aluminum vapour deposition and an additional photolithographic patterning procedure. As a final step, the chip was annealed at 200 °C for 30 min. The surface of the chip was covered by a 30 μm thick layer of polyimide membrane to insulate the transducer electrically from flow liquid. The voltage output per degree of the integrated thermopile was about 2 mV/K at 22 °C.

On the chip, a silicone rubber membrane (0.32 mm thick was used to form the microchannel (17.5 × 3.6 × 0.32 mm) and to serve as a seal between the chip and the plexiglass cover. The inlet and outlet stainless steel tubing, as well as the electrical connectors, were mounted on the cover. The entire unit was held together with a screw-mounted delrin holder. This rather bulky construction was required in order to facilitate repeated access to the sensor chip. The CPG beads were charged into the microchannel by sucking them in from the outlet end. The beads were stopped at the hot junction using a filter made from a tiny piece of kleenex tissue. Two thirds of the channel from the hot junction were filled with the enzyme-containing beads. The remaining third was filled with similar beads without any enzyme, in order to reduce carryover of heat to the cold junction.

2.3.2
Thermistor-Based Microbiosensor

The thermistor-based device (Fig. 6), is composed of a transducer chip (21 × 9 × 0.57 mm), a spacer, and electrical and liquid flow connections. Five thermistors (T_0–T_4) with a temperature coefficient of 1.7% per degree (25 °C)

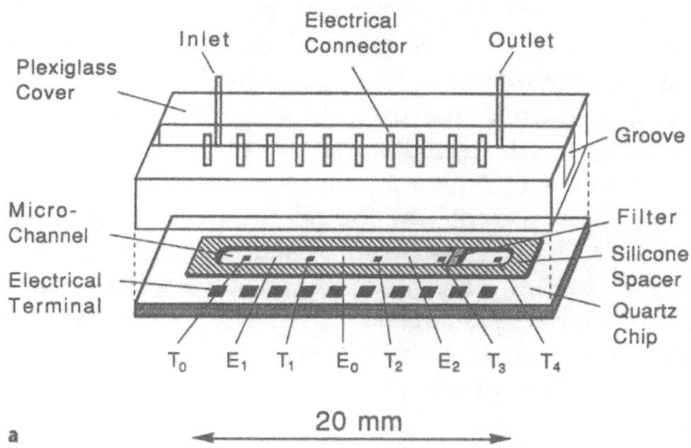

a

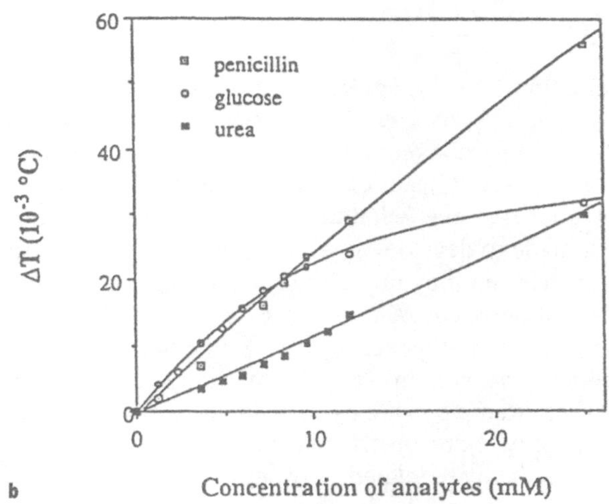

b

Fig. 6. **a** Schematic diagram of the sensor construction. T_0 to T_4 represent the film thermistors 0 to 4, respectively. E_1 and E_2 contain enzyme matrices 1 and 2. E_0 represents the region containing the same carrier beads as the other regions but without immobilized enzymes. **b** Calibration curves for the simultaneous responses of glucose, urea and penicillin in a mixture. Oxygen was electrocatalytically generated in the buffer stream for the glucose oxidase reactions

were fabricated on the quartz chip along the microchannel with a spacing of 3.5 mm by doping in polysilicon and etching.

The thermistors were electrically insulated by depositing a layer of silicon oxide (low temperature oxide). The thermistors were then paired in two independent groups, T_0 with T_1 and T_2 with T_3 corresponding to the enzyme reactions in region E_1 and E_2 respectively. A silicone-rubber membrane (0.32 mm thick) formed the reactive channel ($17.5 \times 0.8 \times 0.32$ mm) and was also used for sealing the device. In order to determine two substrates simultaneously, the enzyme regions E_1 and E_2 were charged with two different enzymes, which had been covalently immobilized on NHS-activated (*N*-hydroxy succinimide) agarose beads (13 µm in diameter, Pharmacia Biotech, Sweden). The E_0 region was charged with similar beads without any enzymes, in order to damp the thermal carryover downstream at region E_2. In this scheme, T_1/T_3 and T_0/T_2 were employed as the measurement- and the reference thermistor, respectively. The agarose beads were held in place, using a filter made of a tiny piece of kleenex tissue. A plexiglas cover, on which the inlet and outlet stainless-steel tubings and electrical connectors were mounted, was used to seal the holder. This design was rather bulky but was required in order to facilitate repeated access to the sensor chip. A typical example of multisensing of penicllin, glucose and urea is shown in Figure 6b.

2.4 Multisensing Devices

An important area in clinical diagnosis is the simultaneous determination of multiple analytes. This approach could be extended to personal healthcare, bioprocess control, and sequential enzyme reactions. In particular, it is essential in the realization of a personal healthcare system, as the information from multiple metabolites improves the reliability of the clinical diagnosis. Immense efforts have been made to develop biosensors for the determination of multiple analytes. Many of these employ multi-channel or split-flow systems, combined with electrochemical detection. A requisite of the multi-channel scheme lies in avoiding interference from other reactions. Nevertheless, uniformity of the flow rate in these multichannel systems (Fig. 7), especially in microchannels, is yet to be improved. Earlier attempts made use of a single flow channel in multianalyte determination. In applications involving electrochemical and optical detection, the system must be suitably regulated, in order to minimize the interference that can arise due to change in pH, ionic strength, electrocatalytic species, or chromophores produced during the reaction. In addition, the specificity of the electrode or the optical detector for the compound being measured is intrinsically dependent on the applied potential or the wavelength. The number of analytes, especially in whole blood, as well as the nature of the detection system, usually govern the detection conditions. Apart from methods in biosensing, multiple discrete samples [20] have also been measured by a centrifugal blood analyzer method based on a rotating sampling distributor. However, the inherent complexity of the latter technique prevents its routine application in delocalized clinical diagnosis.

In the recent past, multianalyte determination has found increased applications, i.e. specific and multiple reactions favor a system that allows the specific determination of each reaction, using the same principal measurement methods, detectors and conditions. In keeping with this idea, a flow injection thermometric method based on an enzyme reaction and an integrated sensor device was proposed for the determination of multiple analytes. In principle the technique relies on the specificity of enzyme catalysis and the universality of

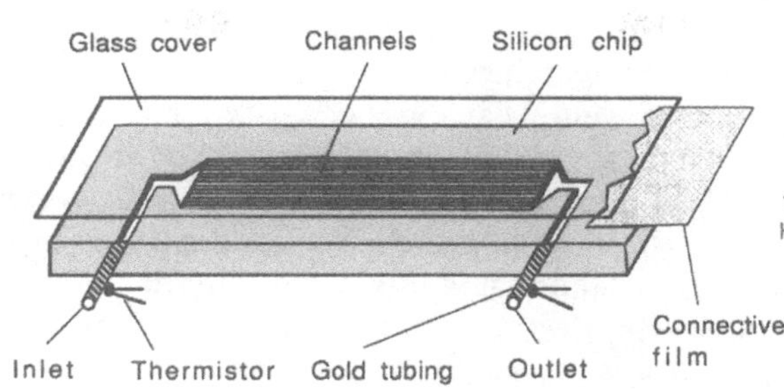

Fig. 7. Schematic illustration of the thermal micro-biosensor fabricated onto a silicon chip

thermal detection. In this technique, a single microchannel column is serially partitioned into several discrete detection regions. Each of the regions, corresponding to the detection of one analyte, contains the corresponding enzyme and a pair of film thermistors. Each thermistor placed after the enzyme matrix, functions as the measurement transducer, whereas the other, placed before the enzyme, serves as the reference. As a substrate mixture flows through this reaction channel, multiple thermal signals generated from individual enzyme reactions are detected nearly simultaneously. One advantage of this design is that all determinations are performed under essentially identical conditions, such as flow rate, sample volume, pressure, and working temperature. Additionally, the effect of by-products of one enzyme reaction on the performance of other enzymes in the series has been found to be minimal.

The feasibility of this approach was demonstrated in dual analytes such as urea/penicillin and urea/glucose. In these investigations each detection region was charged with a different enzyme-agarose bead conjugate (13 µm in diameter). The rest of the channel was charged with a similar bead but without the immobilized enzyme. Complete filling of the channel is necessary to keep down the residence time of the samples within the reactor, as they pass through it. Consequently, determination of multiple analytes could be achieved nearly simultaneously. The error in measurement in such system was primarily caused by thermal carryover. This thermal effect, however, could be minimized by introducing a mini heat-sink between the reaction regions, using silicon or aluminum strips connected with the chip. According to this principle, it would be possible to determine even more analytes by this method, if additional thermal transducers are fabricated. In reactions where different reaction conditions of pH, ionic strength and cofactors are required for the various groups of enzymes/analytes, multi-channels can be supplied. In the case of whole blood and crude samples, large bead size is preferred, in order to prevent clogging and to reduce the back pressure in the flow channel. In addition, direct immobilization of enzymes on a chip surface with extended coupling area would be a better choice if adequate enlargement of surface area could be achieved. It is also possible to employ an integrated thermopile as an alternative to the integrated thermistors in this system. In this case, each thermopile relates to one enzyme. Its one junction (hot junction) is placed downstream of the enzyme matrix to determine the temperature change relative to the other junction (cold junction) maintained upstream at a constant temperature. The advantage of employing thermopiles for the determination of multiple analytes derives from its high rejection ratio of the common-mode thermal noise, and elimination of an additional element for the reference temperature, as is the case in thermistor based sensors.

The integrated system, including transducer and enzyme reactor, provides improved reliability and stability in multianalyte determinations, as compared with discrete thermal sensor systems. In addition, application of micromachining and IC technologies is of benefit for the manufacture of uniform, cheap thermal transducers with flexible shape, size, and resistance, as well as delicate microstructure on the chips. The good thermal insulation of the transducers from the flow stream eliminates interference from the reactants on the transducers, and the intrinsic stability of the transducers obviates the need for frequent recalibration of the sensors.

2.5 Hybrid Biosensors

As the name indicates, hybrid biosensors are an integration of two or more measurement principles for efficient detection of a specific analyte. In general, each type of biosensor has its merits and shortcomings. Electrochemical biosensors demonstrate good selectivity and can be regenerated electrochemically by using electron transfer mediators and/or cofactors. However, while analyzing sample mixtures, electrochemical measurements often suffer from interference by molecules other than the electroactive species being measured. The interference also depends on the applied potential. Although optical techniques have high sensitivity and selectivity, they may be affected by interfering chromophores and fluorophores, present or formed during the reactions. In the case of enzyme-based thermal biosensors, nonspecific heat has to be avoided or balanced by differential measurement. On the other hand, since the thermal transducers are insulated from reactants and buffer, the direct interference with the thermal transducer by chemical compounds in the solution can be eliminated. This enables the determination of complex samples, such as blood using thermal biosensors.

A pioneering hybrid biosensor (Fig. 8a) was designed and demonstrated in our group [21]. This sensor scheme utilizes electrochemical regeneration of the electron mediator in combination with thermal detection, in order to extend the linear range for glucose and catechol measurements (Fig. 8b). Such electrochemical methods had been applied previously in fiber-optic biosensors where the indicator reagent was regenerated electrochemically. In principle, these approaches could be applied for the development of hybrid biosensors based on any oxidase or dehydrogenase. In addition, light-assisted regeneration of NADH may also be of interest in the creation of an optically assisted hybrid biosensor.

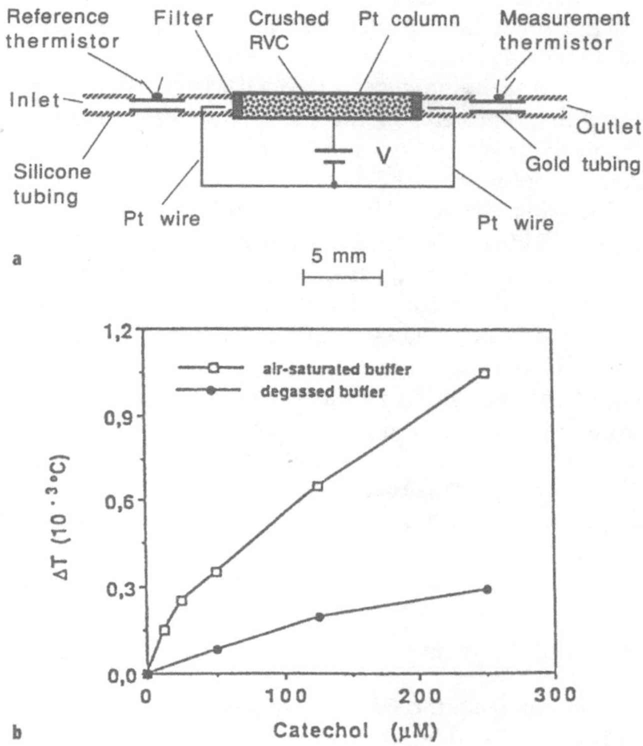

Fig. 8. **a** Schematic of the set-up for simultaneous electrochemical and thermometric determination of analytes (for a detailed explanation, see text). **b** Optimization of catechol detection using hybrid (electrochemical and thermometric sensing) for oxygen concentration and its effect on sensitivity of thermal detection [30]

3
Applications

3.1
General Applications

The metabolites determined using ET devices include alcohols, glucose, and lactate using alcohol oxidase [22], glucose oxidase [23] and lactate oxidase respectively. The detection limit for these compounds in pure solutions was in the submicromolar region. The normal procedure was to co-immobilize the oxidases with catalase in order to increase the total heat production, reduce the oxygen consumption, and eliminate the hydrogen peroxide. Cellobiose was measured by using β-glucosidase in combination with glucose oxidase and catalase. Determination of cholesterol and cholesterol esters was accomplished using cholesterol oxidase and cholesterol esterase, and of triglycerides by using lipase.

Typical sample matrices are blood, serum and fermentation broth. Oxalate in urine was also measured using oxalate oxidase [24]. Substrate recycling offered routes for highly sensitive measurements. An example was lactate (or pyruvate) determination with a co-immobilized LDH/LOD (lactate dehydrogenase/lactate oxidase) column that repeatedly oxidizes lactate to pyruvate and reduces pyruvate to lactate. Each cycle produces a considerable amount of heat [15]. A similar approach was employed for the determination of NAD^+/NADH by co-enzyme recycling by using lactate dehydrogenase plus glucose-6-phosphate dehydrogenase, and of ATP/ADP by coupling pyruvate kinase with hexokinase as the recycling enzymes. By operating at excessive glucose supply, a hexokinase column was used for indirect assay of ATP with micromolar sensitivity. A multiplicative effect could be attained by coupling the recycling systems of pyruvate kinase and of LDH/LOD [25].

Monitoring of specific proteins eluted from chromatographic columns was demonstrated using the ET as a direct online monitor for purification of proteins/enzymes. As an example, LDH was recovered from a solution by affinity binding of N_6-(6-aminohexyl)-AMP-Sepharose gel, and the signal from the ET was used to regulate the addition of the AMP-Sepharose suspension to the LDH solution [26, 27].

3.2 Industrial and Process Monitoring

For bioprocess monitoring, the ET was employed in the assay of penicillin in fermentation broth, using β-lactamase or penicillin acylase [28]. Also, ethanol generated in alcohol fermentation by yeast was monitored using alcohol oxidase. Other metabolites that were monitored during fermentations include lactate, glycerol, acetaldehyde, sucrose and glutamine [29].

3.3 Clinical Applications

Metabolites in human blood are closely associated with the state of an individual's health. Determination of metabolites is critical in clinical diagnosis, since they can serve as a criterion for judging the severity of the sickness. Of these metabolites, glucose, lactate and urea are most frequently determined. Special attention to determination of glucose in blood is due to the fact that diabetes is well known as a dangerous and widespread disease that results in a high glucose concentration in the blood. Some other effects of metabolites, such as urea and lactate, on shock, respiratory insufficiency, and heart and kidney diseases are understood to some extent.

3.3.1 *In-Vitro Monitoring*

For determination of the metabolites in human blood, samples were collected from veins into heparinized or EDTA containing tubes. For glucose analysis, NaF

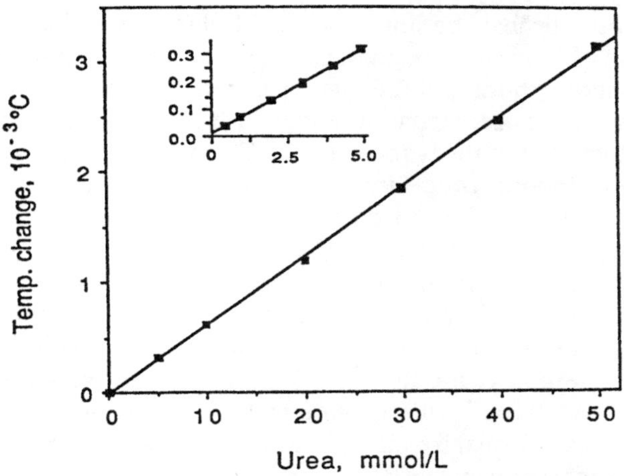

Fig. 9. The linear range for urea detection using a miniaturized thermometric system

preservative was also included in the tubes. For lactate, this preservative does not stabilize the concentration and may inhibit enzyme activity [21, 30]. Urea (Fig. 9) was relatively stable for several hours after withdrawal. Using the thermal biosensors, the blood samples were directly analyzed without any additional treatment. The sampling rate in this case was from 12 to 90 assays per hour. Meanwhile, the same samples were determined with the reference methods. In the UV reference method, it is necessary to deproteinize the blood in order to stop the metabolism of several unstable analytes, since this method takes much longer than the biosensor method. This was achieved by adding perchloric acid to the blood at the start of the thermometric determination, in order to reduce the measurement error due to the concentration change in the sample. Higher concentrations of blood metabolites for comparison were achieved by adding small volumes of high concentration standards into the native blood samples. The results of the determination of glucose, urea and lactate [31] in whole blood are summarized in Table 2.

Table 2. Determination of glucose, urea, and lactate in undiluted blood using miniaturized thermal sensors

Metabolite	Linear range (mM)	CV (%)	Sample volume (μl)	Correlation coefficient	Reference method
Glucose	0.5–20	3.7 [a]	1	0.980 [c]	Reflolux-S
Urea	0.2–50	4.1 [b]	1	0.989 [d]	UV
Lactate	0.2–14		1	0.984 [d]	UV

[a] For 100 samples, [b] For 50 samples, [c] For 37 samples, [d] For 30 samples.

The results indicate that the linear ranges of glucose and lactate (oxidase reactions) in whole blood correspond well with those of the same metabolites in buffers. The sensor methods and the reference methods were in good correlation for all analytes. The precision for the standards in buffers was always better (about 2 – 3%) than that of the blood samples. This was ascribed in part to the instability of the metabolite concentrations in blood, particularly that of lactate. In addition, the blood viscosity and the nonspecific heat in the reaction can also affect the final results.

The primary features of this method are its general principle, a uniform measurement system, the use of untreated blood samples, minute sample volumes, no fouling of the transducers, no electrochemical or optical interference, simple procedures, rapid response, and low cost. According to the working principle, other metabolites associated with enzyme reactions – in addition to glucose, urea and lactate – can be analyzed in a similar way. Unlike electrochemical and optical detection, where the potential or wavelength must be adapted to a specific analyte, no modification is needed for this measurement system other than replacement of the enzyme matrix or column. Isolation of the thermal transducers from the reactants avoids fouling from blood samples, and facilitates the stability and long-term operation of the sensors. The requirement of small amounts of enzyme and sample, as well as the capability of more than 100 blood assays per enzyme column, made the determination cheap and convenient, for instance by using capillary blood taken from the finger. Recently cholesterol determination was revisited, and a procedure was developed that allows estimation of free and esterified LDL/HDL cholesterol [32].

A miniaturized thermal biosensor was evaluated [33] as part of a flow-injection analysis system for the determination of glucose in whole blood. Glucose was determined by measuring the heat evolved when samples containing glucose passed through a small column containing immobilized glucose oxidase and catalase. Samples of whole blood (1 μl) were measured directly, without any pretreatment. The correlation between the response of the thermal biosensor and other devices, i.e. the portable Reflolux S meter (Boehringer Mannheim, Fig. 10a), the colorimetric Granutest 100 glucose test kit (Merck Diagnostica) and the Ektachem (Kodak) instrument (Fig. 10b), was evaluated. The influence of the hematocrit value and of possible interference was reported. The correlation measurements showed that the thermal biosensor generally give lower values than the reference methods when aqueous buffer standards were made for calibration of the ET. Mean negative biases range from 0.53 to 1.16 mM. Differences in sample treatment clearly complicated the comparisons and the proper choice of reference method. There was no influence from substances such as ascorbic acid (0.11 mM), uric acid (0.48 mM), urea (4.3 mM) and acetaminophen (0.17 mM), on the response to 5 mM glucose. The hematocrit value did not influence the glucose determination, for hematocrit values between 13 and 53%.

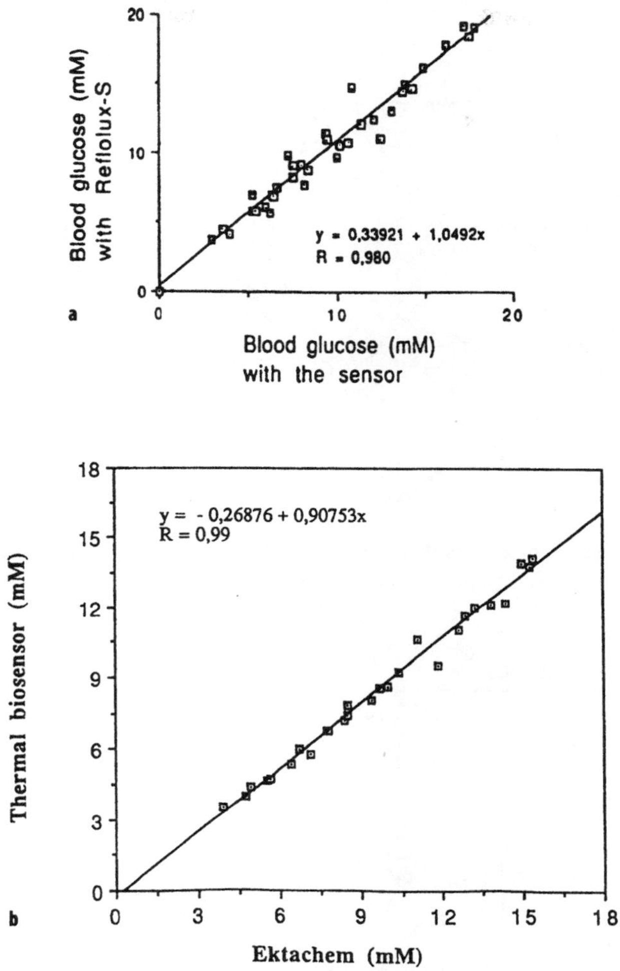

Fig. 10. **a** Correlation of glucose measurement in whole blood, measured by micro-thermometric biosensor and by the Reflolux-S blood glucose analyser. **b** Correlation of glucose measurement in whole blood measured by micro-thermometric biosensor and by the Ektachem blood glucose analyzer

3.3.2
In-Vivo Monitoring

A miniaturized thermal flow-injection analysis biosensor was coupled with a microdialysis probe for continuous subcutaneous monitoring of glucose [34]. The system (Scheme 1) consisted of a miniaturized thermal biosensor with a small column containing co-immobilized glucose oxidase and catalase. The analysis buffer passed through the column at a flow rate of 60 µl/min via a 1-µl sample loop connected to a microdialysis probe (Fig. 11).

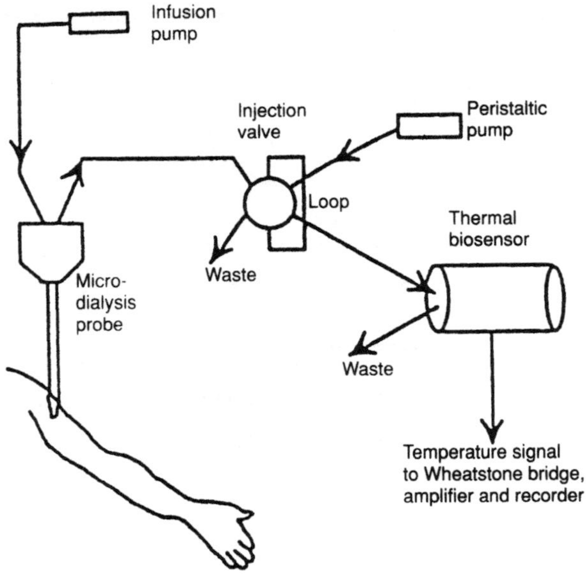

Scheme 1. Setup for simultaneous thermometric and blood-glucose analyser measurements, using a micro-dialysis in-vivo probe on a human volunteer

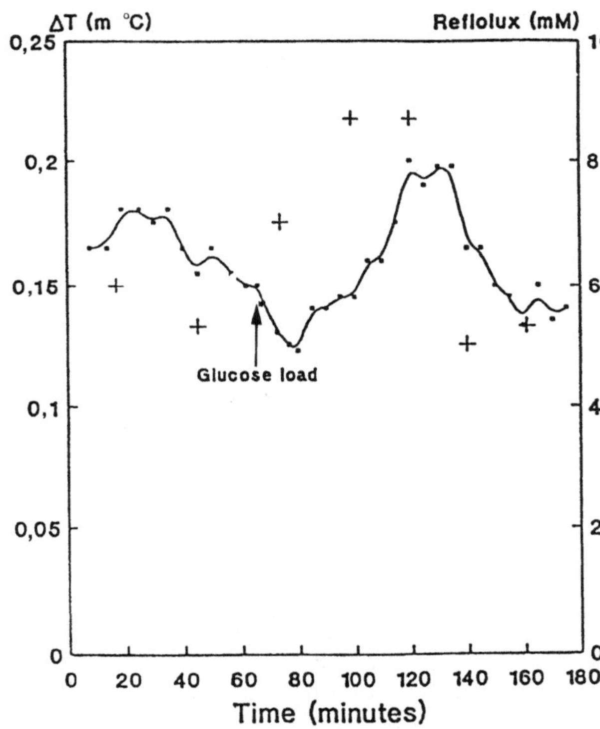

Fig. 11. In-vivo glucose load with a healthy volunteer. Points plotted on the graph (+) are the glucose concentrations measured with a blood glucose analyser and (–) are the peaks of temperature change. The dialysis fiber (10 mm) was inserted subcutaneously. The perfusion rate was 3 μl/min

3.3.3
Bedside Monitoring

A prototype for a bedside monitoring system was developed for semi-continuous monitoring of blood-glucose concentration, requiring only one calibration point [35]. This was made possible by using the special advantage of the thermometric sensing technique in combination with the adjustment of flow. The glucose concentration was determined from the difference between the sensor response and an estimated background signal. Using standard addition technique, calibration factors for background and sensitivity were set and remained unchanged during the monitoring. Recovery in whole blood was 90–98% with an injection interval of 3 min and the precision of the sensor was <3% over more than 100 blood samples. Response time was about 60 s. The calculated glucose values correlated well with a YSI glucose analyser over a range of 2–20 mmol/l.

3.3.4
Multianalyte Determination

Multiple analytes were determined simultaneously by a flow injection thermal micro-biosensor. The biosensor consisted of five or more thin film thermistors located along a single microchannel. The device was fabricated on a quartz chip by micromachining. The feasibility of employing this system for the detection of two independent enzyme reactions was demonstrated using two different pairs of enzymes, urease/penicillinase and urease/glucose oxidase [37]. The enzymes were immobilized on agarose beads which were then sequentially packed into distinct regions of the microchannel. Using this method, samples containing urea mixed with penicillin-V or with glucose were determined simultaneously. The sensor was capable of analysing 25 samples/hour. This study was followed by three and four analyte measurements [19].

3.3.5
Hybrid Sensors; Enzyme Substrate Recycling

A combination of calorimetric and electrochemical detection principles led to the creation of a novel biosensor [38] that retained the principal advantages of both techniques. In order to demonstrate the feasibility of such an approach, a ferrocene-mediated thermal flow-injection glucose sensor was fabricated and tested (see Scheme 2). The electrochemical reaction was accomplished by applying a voltage between a platinum column (working electrode), in contact with a crushed reticulated vitreous carbon RVC matrix onto which glucose oxidase was immobilized; and platinum wires (counter electrode) were located at the inlet and outlet of the column. For detection, the thermal signal generated by the glucose oxidation reaction was measured in conjunction with the electrochemical signal. By using this method, a linear range of glucose concentration upto 20 mmol/l was achieved, independent of the oxygen concentration in the buffer.

In a similar approach catechol was measured using tyrosinase [30]. The enzyme column was constructed of a platinum foil in electrical contact with a

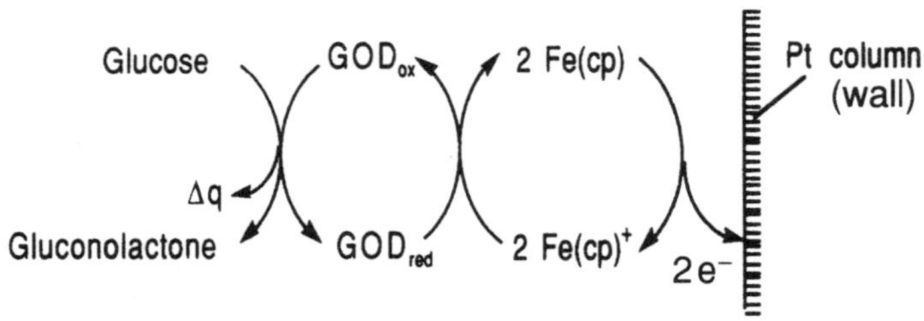

Scheme 2. Principle of the ferrocene-mediated thermal glucose biosensor, Δq indicates the heat produced in the reaction

poly(pyrrole)-coated reticulated vitreous carbon (RVC) matrix onto which tyrosinase was immobilized. The column functioned as enzyme reactor, working electrode and thermally sensitive element together with the thermistors. Catechol was oxidized by tyrosinase to form 1,2-benzoquinone, which was subsequently regenerated electrochemically on the electrode surface. The primary heat production developed by the enzyme reaction could be measured calorimetrically. Simultaneously, the electrochemical reduction of 1,2-benzoquinone generated a current that was detected by the working electrode. Such hybrid sensors provide a useful tool for comparative studies of complex reaction schemes.

3.4 Other Applications

3.4.1 *Enzyme-Activity Measurement*

Significant interest has been generated in monitoring enzyme activity and metabolite concentration in non-aqueous solvents [39, 40]. For example, the reactions of immobilized lipoprotein lipase with tributyrin in a buffer-detergent system and cyclohexane were compared using an ET. Also it was demonstrated that horseradish peroxidase produced a considerably higher temperature signal in toluene than in water. Furthermore, addition of diethyl ether in small amounts was found to enhance this effect. In an analogous approach, the reaction of chymotrypsin in 10% DMF for hydrolysis (exothermic) and synthesis (endothermic) of peptide bonds was monitored using the ET.

3.4.2 *Food*

Several metabolites found in food samples have been estimated using the thermometric approach. These include glucose, cellobiose, lactose, maltose, galactose, lactate, oxalate, phopholipids, ascorbic acid, ethanol, urea [18], xanthine and hypoxanthine [41, 42]. Glucose was estimated by co-immobilizing glucose oxidase and catalase [43]. The presence of catalase doubled the thermal response as it restored half the oxygen consumed in the GOX reaction. It also eliminated the H_2O_2 formed in the GOX reaction. Linearity up to 0.7–1.0 mM could be achieved by this technique. The limit of detection was about 1 µM. Glucose in blood plasma and serum were determined by this method [3]. The use of this system has also been extended to the detection of glucose in hydrolysates of cellobiose, lactose and maltose. This approach was advantageous, as low enthalpy changes made it difficult to monitor the hydrolysis directly due to low enthalpy changes [44–46].

Employing L-ascorbate oxidase, vitamin C (ascrobic acid) was determined in food samples between 0.05–0.6 mM [44]. In order to measure ethanol in beverages and blood samples, alcohol oxidase from *Candida boidinii* had been employed. Linearity was obtained between 0.01 and 1 mM. These measurements were also found to be useful in monitoring fermentation [47, 48].

L-lactate and oxalate were also tested with lactate-2-monoxygenase and oxalate decarboxylase and excellent results were obtained. CPG columns were employed in both instances. Good linearity was obtained between 0.005–1 mM for L-lactate [3] and between 0.1–3 mM for oxalate [24]. Similarly, urea was measured with a precision better than 1% in the linearity range 0.01–200 mM using Jack bean urease. The reaction of urea with ethanol to produce ethylcarbamate is of interest in fermentation monitoring.

Lipids such as triglycerides were determined with lipoprotein lipase and phospholipids with phospholipase D. In the case of triglycerides, good linearity between 0.05–10 mM (tributyrin) and 0.1–5 mM (triolein) was obtained, whereas for phospholipids linearity was obtained between 0.03–0.19 mM [41].

3.4.3
Environmental

Two different concepts were employed for this purpose: substance-specific analysis using enzymes (substrate or inhibition) and more general measurements applying whole cells. ET was successfully applied [49] to the monitoring of heavy metal (Hg^{2+}, Cu^{+2} and Ag^{+}) toxicity in the environment by measuring the inhibition of urease activity down to ppb levels of the metal ions. Restoration of activity was also tested upon chelation of the metal ions with strong chelating agents. In the recent past, a study of Cu(II) determination was carried out using acid urease [50]. In addition, Satoh et al. [51, 52] described flow injection microdeterminations using enzyme thermistors with different immobilized enzymes for the detection of heavy metal ions. The heavy metal ions were detected due to their reactivating effect on apoenzymes.

In another configuration [53] two different approaches for pesticide analysis were employed. A crude enzyme solution capable of hydrolyzing organophosphate insecticides was prepared. The enzyme was coupled to controlled-pore glass with glutaraldehyde. The insecticides, e.g., parathion, cyanophos, and diazinon, were dissolved in a perfusion buffer (Tris pH 8.9, 1% Triton X-100) and injected as a 10 min pulse into an ET in split-flow mode. The instrument measured the heat output due to insecticide hydrolysis and consecutive buffer protonization. For parathion, the detection limit was approximately 10 ppm.

The second approach [54] was based on the inhibition of acetylcholine esterase. One unit of acetylcholine esterase was reversibly immobilized via lectin binding to Con A-Sepharose and could be rinsed off with a pulse of 0.2 M glycine-HCl, pH 2.2. Reversible immobilization of enzymes and whole cells in the enzyme thermistor column, utilising specific lectin-glucoprotein interactions, had been introduced earlier and was especially useful for inhibition studies, where the enzyme had to be replaced very often. Enzyme activity was determined with 10 mM butyrylcholine as a substrate. A 5 – 10 min pulse of pesticide solution was introduced into the flow buffer, followed by a second substrate pulse. The decrease in activity was proportional to the amount of pesticide, with a detection limit below 1 ppm.

In order to adapt the system to on-line monitoring, in wastewater control, for example, the occurrence of pesticide in a flow buffer was investigated. It was found possible to differentiate between reversible and irreversible inhibition and to quantify a reversible inhibitor. Since it was possible with the calorimetric method to use the natural substrate acetylcholine to assay cholinesterase, instead of the commoly used thiocholines, this methodology might be useful in medical research as well.

Whole cells were employed [55] as the monitoring element in which *Pseudomonas capacia* capable of metabolizing aromatic compounds were immobilized in Ca^{+2}-alginate beads and their response to aromatic substances, e.g., salicylate, was monitored with an enzyme thermistor.

3.4.4
Fluoride Sensing

More recently, it was demonstrated that the thermistor approach could be used to monitor specific interactions of fluoride ions with silica-packed columns in the flow injection mode. A thermometric method for detection of fluoride [56] was developed that relies on the specific interaction of fluoride with hydroxyapatite. The detection principle is based on the measurement of the enthalpy change upon adsorption of fluoride onto ceramic hydroxyapatite, by temperature monitoring with a thermistor-based flow injection calorimeter. The detection limit for fluoride was 0.1 ppm, which is in the same range as that of a commercial ion-selective electrode. The method could be applied to fluoride in aqueous solution as well as in cosmetic preparations. The system yielded highly reproducible results over at least 6 months, without the need of replacing or regenerating the ceramic hydroxyapatite column. The ease of operation of thermal sensing and the ability to couple the system to flow injection analysis provided a versatile, low-cost, and rapid detection method for fluoride.

3.4.5
Cellular Metabolism

The effects of ampicillin-induced spheroplast formation on the production of molecular hydrogen by *Escherichia coli* carrying out fermentation in a lactose-peptone broth with an osmolality of 342 mosmol/l was investigated previously

[57]. The effects were most pronounced during the transformation of bacterial cells to spheroplasts. It was shown that the lower production rate of molecular hydrogen by spheroplastic cells was due not only to a suggested decrease in mixed-acid fermentation, but to a reduction in hydrogen lyase activity as well.

The production of molecular hydrogen was measured in the effluent gas of seven fermentations [58]. The aim of this primary investigation was to study the use of a H_2-sensitive metal-oxide-semiconductor structure in physiological studies of *Escherichia coli*. In order to yield more information, the metabolic heat was measured with a flow microcalorimeter in parallel with the determination of molecular hydrogen.

3.5
Miscellaneous Applications

The characterization of immobilized invertase was carried out, and the technique was successfully coupled to the catalytic activity determination of immobilized cells [59]. Similarly, the results of this technique were useful in the selection of *Trigonopsis variabilis* strains for high cephalosporin-transforming activity [60]. Also, the cephalosporin-transforming activity of D-amino acid oxidase isolated from yeast was identified in a similar manner. The thermometric signal was proportional to the number of cells as well as the amount of D-amino acid oxidase immobilized in the ET microcolumn. The ET was also coupled to a thermometric ELISA procedure (TELISA) for the determination of hormones, antibodies and other biomolecules generated during the fermentation process [61]. Genetically engineered enzyme conjugates, e.g. human proinsulin-alkaline phosphatase conjugate, were used for the determination of insulin or proinsulin. Alkaline phosphatase was used predominantly as the enzyme label for such an assay [62]. In another instance TELISA was employed for monitoring insulin separation [63]. The expense of the conjugate for such automated procedures was found to be negligible compared to the higher costs of the non-automated procedures. In addition these techniques are easily set-up in an industrial environment and have already been tested in several instances, e.g. monitoring of fermentation.

Enzyme thermistors have also found applications in more research-related topics, such as the direct estimation of the intrinsic kinetics of immobilized biocatalysts [64]. Here, the enzyme thermistor offered a rapid and direct method for the determination of kinetic constants (K_i, K_m and V_m) for immobilized enzymes. For the system being investigated, saccharose and immobilized invertase, the results obtained with the enzyme thermistor and with an independent differential reactor system were in very good correlation, within a flow-rate range of 1 to 1.5 ml/min.

Determination of ADP and ATP by multiple enzymes in recycling systems: pyruvate kinase and hexokinase co-immobilized on aminopropyl CPG, was demonstrated by Kirstein et al. [25]. In addition, a second reactor,with L-lactate dehydrogenase, lactate oxidase, and catalase, was used to increase the sensitivity from 6×10^{-5} M with no recycling, to 2×10^{-6} M in the kinase bienzyme reactor, and – finally – to 1×10^{-8} M with the dual recycling system, corresponding to an overall 1700-fold multiplication.

The use of an enzyme thermistor as a specific detector for monitoring different enzymes [17, 63, 45] in the eluents from chromatographic procedures had the advantage of being applicable in optically dense solutions, where spectrophotometric methods fail, and of being able to operate on-line for discrete samples.

In 1988, Flygare et al. [39] made use of the ability of enzymes to function as catalysts in organic solvents. Performing the biosensor analysis in these solvents with improved substrate and product solubilities sometimes changed the substrate specificity of the enzyme to a specific substrate, or even led to new by-products. A specific advantage for thermal analysis in organic solvents was their lower heat capacity and higher thermal expansion coefficient, leading to a large gain in sensitivity [65].

Flygare et al. [40] also used the enzyme thermistor for the control of an affinity purification. Here, lactate dehydrogenase (LDH) was recovered from a solution by binding to a special Sepharose gel (AMT-sepharose). The addition of the gel to the solution was controlled by a PID controller or a desktop computer, according to the amount of unbound LDH detected with the enzyme thermistor. Both systems enabled rapid and accurate assessment of the correct addition of the adsorbent.

4 Future Developments

4.1 Telemedicine

The rapid advances in microelectronic technology have resulted in revolutionizing the field of information technology. Today, computers and communication facilities offer a number of possibilities for upgrading our everyday life and work. It may also affect our life with respect to healthcare. Such advances in information technology can be intelligently utilized in tele-medical monitoring and in diagnostic systems for decentralized use.

These systems would be able to fulfil functions, such as health monitoring, routine check-up, clinical diagnosis, and to provide medical advice. They could take over a role in the primary healthcare of humans and make it convenient for medical practitioners to assess and treat the diseases instantly. As patients can be easily linked to a central diagnostic and monitoring facility through the telecommunication network, they can avail themselves of it either at home or during travel. This technology is of strategic importance, as it can reduce the cost of healthcare, and improve the quality of life by creating a more secure and healthy environment.

This technology has been launched in some countries, where physiological parameters are at present being collected by physicians from remote locations. In general, telemedicine is a broad concept that includes not only transfer of medical documents, such as X-ray film, but – more importantly – connection between the human body with the physicians through a "biomedical interface" that collects the biomedical information from the body for data processing and communication. For healthcare in the home, however, the tele-medical system, relies extensively on the development of an "interface" between the human body and the computer/communication facilities. In contrast to the highly developed information processing and transfer technology and established clinical diagnostic criteria, the acquisition of biomedical information (physiological and biochemical parameters) is an active area being pursued in our group.

As compared with the electrochemical and optical biosensors mentioned above, thermal biosensors are intrinsically insensitive to the optical and electrochemical properties of the samples, and do not require frequent calibration of the transducers, since the transducers are highly stable and are normally isolated from the buffer and sample fluids. Recently, highly sensitive integrated thermal biosensors have been developed in our group. As discussed above, they employ micromachining and semiconductor technology as well as control software of computer data for simultaneous determination of multianalytes (up to four) in mixed samples [19]. Glucose, lactate and urea in1 μL whole blood samples could be directly determined with a miniaturized sensor without any pretreatment of the samples. These achievements show promise for further development of a fully integrated analytical system based on a thermal biosensor array and microchannel fluid handling. This system will eventually be incorporated with computer and telecommunication facilities to accomplish the telemedical monitoring and diagnostic system for home healthcare.

4.2
Home Diagnostics

Immense progress in the field of glucose analysis, which is of special significance for diabetes patients, was achieved with integrated silicon thermopiles, miniaturised enzyme thermistors, and new calorimetric microbiosensors. Some years ago, our group [36] started development of microbiosensors that could be produced by micromachining. As an intermediary step, miniaturised enzyme thermistor models were produced which were found to perform unexpectedly well, in spite of their relatively simple design. With respect to sensitivity, precision, physical dimensions, and column longevity (cost per assay), these devices were quite competitive with instruments currently being used for home measurements of blood glucose in diabetics. This could be improved provided a suitable pump and sample injection valve could be developed [21]. Carefully selected enzyme support materials made it possible to run untreated whole-blood samples directly through the immobilised enzyme column.

4.3
Other Developments

The feasibility of miniaturizing thermal biosensors in different constructions, sizes and materials, by employing conventional machining and micromachining technologies could be further exploited in several directions. The miniaturization has led to the improvement of the sensor's sensitivity and response time. These developments are promising for the direct analysis of physiological samples, e.g. undiluted whole blood. The feasibility of microreactor fabrication on the surface of silicon coupled to thermal detection using immobilized enzymes is also trend-setting. The efficiency in such systems was much better when the amount of immobilized enzyme was compared to the gain in sensitivity. In addition, integration of thin-film thermistors/ thermopiles permitted simultaneous determination of multiple analytes. At present, four separate analytes have been discriminated, and additional efforts to extend the range to the detection of several analytes is in progress. Furthermore, the suitability of the construction for small volume measurements, i.e. < 1 µl, has resulted in an increased linear range, for example in glucose sensing with glucose oxidase.

Application of the miniaturized biosensors for metabolite estimation in whole blood is another important concept for future development. The improvement in sensitivity, linear range and response time was achieved by miniaturization of the sensors and has been proven in the case of whole-blood glucose, urea and lactate. A useful feature of miniaturization is that the smaller the flow channel, the smaller are the dispersion and dilution effects, favouring whole blood measurement with minimum error. In the case of clinical estimations, the results of the measurement on miniaturized thermal biosensor are more reliable.

The concept of "the home doctor" is a multi-disciplinary project integrating expertise from several areas of scientific research. It would include the developments in miniaturized biosensors coupled to communication technology. Additional help would be essential from computer scientists and clinical chemists. In this regard it would be imperative to enhance the sensitivity of thermal biosensors in order to make them more reliable and reproducible. This factor is mainly governed by improvement in integrated circuit technology and reduction of the heat capacity of reactors. In addition, for such devices, optimizing the doping concentration or the manufacturing process would improve the signal-to-noise ratio of the thermistors. Moreover, the output of thermopiles could be enhanced by integration of several thermocouples, especially in the case of multiple metabolite determination which requires larger transducer arrays to decrease the thermal interference. This could be achieved by introducing a micro-heat sink constructed of silicon or aluminium into the device, thus immensely decreasing the thermal carry-over. Integrating the amplifier with the thermopile on chips would further improve the stability of the thermal signal. Diaphragm structures fabricated on crystal chips would also further improve the reactor heat capacity.

Metabolites other than glucose, urea and lactate, can be monitored using the thermometric technique. In the case of metabolites that cannot be estimated enzymatically, other techniques, such as electrochemical and optical methods, would have to be integrated in tandem with thermal sensing. Following thermometric sensing, the clinical interpretation of the results is equally important. This could be accomplished using data-processing systems, according to which the clinical status and the correlation between the metabolites and various disease states could be evaluated.

A hybrid biosensor combining principles of electrochemistry, flow injection analysis, and calorimetry has also been proposed and developed by us. The concept *per se* is extremely vital and could be extended to other combinations such as opto-thermal, etc. An algorithm for other forms of hybrid biosensors is presently being designed within the group, essentially aimed at extending the range of detectable analytes in combination with improved sensitivity and linear range. The impact of hybrid biosensors, created by interdisciplinary cooperation would have an immense potential on the developmental success of biosensor applications. As it has been already demonstrated in the case of thermal and electrochemical biosensors, it should be possible to extend this application to other oxidases and dehydrogenases. In addition, this concept is of interest for developing other hybrid biosensors, such as photoassisted biosensors (light regeneration of NAD^+) wherein different configurations of the reaction cell for photochemical reactions could be employed in conjunction with thermometric sensing.

5 Conclusions

Based on the plethora of applications of thermistor/thermopile based devices, it can be concluded that the field of thermometric sensing offers several avenues of progress in materials science, process monitoring, process control, molecular level detection, characterization of biocatalysts, hybrid sensing and multisensing devices, as well as in telemedicine and other areas of biomedical analysis.

6 References

1. Spink C, Wadsö I (1976) Meth Biochem Anal 23:1
2. Danielsson B, Mosbach K (1986) In: Turner APF, Karube I, Wilson GS (eds) Biosensors: Fundamentals and Applications. Oxford University Press, Oxford, p 575
3. Danielsson B, Mosbach K (1988) Methods Enzymol 137:181
4. Mosbach K, Danielsson B (1974) Biochim Biophys Acta 364:140
5. Weaver JC, Cooney CL, Fulton SP, Schuler D, Tannenbaum SR (1976) Biochim Biophys Acta 452:285
6. Pennington SN (1976) Anal Biochem 72:230
7. Tran-Minh C, Vallin D (1978) Anal Chem 50:1874
8. Rich S, Ianiello RM, Jesperson ND (1979) Anal Chem 51(2):204
9. Bowers LD, Carr PW (1976) Clin Chem 22:1427
10. Schmidt H-L, Krisam G, Grenner G (1976) Biochim Biophys Acta 429:283
11. Kiba N, Tomiyasu T, Furusawa M (1984) Talanta 31:131
12. Muramatsu H, Dicks JM, Karube I (1987) Anal Chim Acta 197:347
13. Muehlbauer MJ, Guilbeau EJ, Towe BC (1989) Anal Chem 61:77
14. Urban G, Jachimowicz A, Kohl F, Kuttner H, Olcaytug F, Goiser P, Prohaska O (1990) Sens and Actuators A21-A23:650
15. Scheller F, Siegbahn N, Danielsson B, Mosbach K (1985) Anal Chem 57:1740
16. Mattiasson B, Danielsson B, Mosbach K (1976) Anal. Lett 9:867
17. Danielsson B, Mattiasson B, Mosbach K (1981) Appl Biochem Bioeng 3:97
18. Danielsson B, Gadd K, Mattiasson B, Mosbach K (1976) Anal Lett 9:987

19. Xie B, Mecklenburg M, Dzgoev A, Danielsson B (1996) Analytical Methods and Instrumentation, special issue, μTAS '96:95
20. Schembi CT, Ostoich V, Lingane PJ, Burd TL, Buhl SN (1992) Clin Chem 38:1665
21. Xie B, Hedberg U, Mecklenburg M, Danielsson B (1993a) Sens Actuators B15–16:141
22. Guilbault GG, Danielsson B, Mandenius CF, Mosbach K (1983) Anal Chem 55:1582
23. Mandenius CF, Buelow L, Danielsson B, Mosbach K (1985) Appl Microbiol Biotechnol 21:135
24. Winquist F, Danielsson B, Malpote J-Y, Larsson M-B (1985) Anal Lett 18:573
25. Kirstein D, Danielsson B, Scheller F, Mosbach K (1989) Biosensors Bioelectronics 8:205
26. Danielsson B, Mosbach K (1979) FEBS Lett 101:47
27. Danielsson B, Rieke R, Mattiasson B, Winquist F, Mosbach K (1981) Appl Biochem Biotechnol 6:207
28. Decristoforo G, Danielsson B (1984) Anal Chem 56:263
29. Rank M, Gram J, Stern-Nielsson K, Danielsson B (1995) Appl Microbiol Biotechnol 42(6):813
30. Xie B, Tang X, Wollenberger U, Johansson G, Gorton L, Scheller F, Danielsson B (1997) Anal Lett 30(12):2141
31. Xie B, Harborn U, Mecklenburg M, Danielsson B (1994) Clin Chem 40(12):2282
32. Raghavan V, Ramanathan K, Sundaram PV, Danielsson B (1998) (submitted)
33. Harborn U, Xie B, Venkatesh R, Danielsson B (1997) Clin Chim Acta 267:225
34. Amine A, Digua K, Xie B, Danielsson B (1995) Anal Lett 28(13):2275
35. Carlsson T, Adamson U, Lins P-E, Danielsson B (1996) Clin Chim Acta 251:187
36. Xie B, Danielsson B, Norberg P, Winquist F, Lundström I (1992) Sens Actuators B6: 127
37. Xie B, Mecklenburg M, Danielsson B, Öhman O, Norlin P, Winquist F (1995) Analyst 120:155
38. Xie B, Khayyami M, Nwosu T, Larsson P-O, Danielsson B (1993b) Analyst 118:845
39. Flygare L, Danielsson B (1988) Ann NY Acad Sci 542:485
40. Flygare L, Larsson P-O, Danielsson B (1990) Biotechnol Bioeng 36:723
41. Satoh I (1988) Meth Enzymol 137:217
42. Satoh I, Inoue K, Arakawa S (1988) Tech. Digest, 7th Sensor Symposium Tokyo, p 229
43. Danielsson B, Gadd K, Mattiasson B, Mosbach K (1977) Clin Chim Acta 81:163
44. Mattiasson B, Danielsson B (1982) Carbohydr Res 102:273
45. Danielsson B, Buelow L, Lowe CR, Satoh I, Mosbach K (1981) Anal Biochem 117:84
46. Mandenius CF, Danielsson B (1988) Meth Enzymol 137:307
47. Danielsson B (1991) In: Blum LJ, Coulet PR (eds) Biosensor Principles and Applications. Marcel Dekker, New York, p 83
48. Rank M, Gram J, Danielsson B (1992) Biosensors and Bioelectronics 7:631
49. Mattiasson B, Danielsson B, Hermannsson C, Mosbach K (1978) FEBS Lett 85(2):203
50. Preininger C, Danielsson B (1996) Analyst 121:1717
51. Satoh I (1991) Netsu Sokutei 18(2):89
52. Satoh I (1992) Ann NY Acad Sci 672:240
53. Mattiasson B, Rieke E, Munneke D, Mosbach K (1979) J Solid-Phase Biochem 4(4):263
54. Mattiasson B, Borrebaeck C (1978) FEBS Lett 85(1):119
55. Thavarungkul P, Håkanson H, Mattiasson B (1991) Anal Chim Acta 249:17
56. Salman S, Haupt K, Ramanathan K, Danielsson B (1997) Anal Comm 34:329
57. Hörnsten EG, Nilsson LE, Danielsson B (1990) Appl Microbiol Biotechnol 32:455
58. Hörnsten EG, Danielsson B, Elwing H, Lundström I (1986) Appl Microbiol Biotechnol 24:117
59. Gemeiner P, Docolomansky P, Nahalka J, Stefuca V, Danielsson B (1996) Biotech Bioeng 49:26
60. Gemeiner P, Stefuca V, Welwardova A, Michalkova E, Welward L, Kurillova L, Danielsson B (1993) Enzyme Microb Technol 15:50
61. Birnbaum S, Buelow L, Hardy K, Danielsson B, Mosbach K (1986) Anal Biochem 158: 12

62. Mecklenburg M, Lindbladh C, Li H, Mosbach K, Danielsson B (1993) Anal Biochem 212:388
63. Danielsson B, Larsson P-O (1990) Trends Anal Chem 9(7):223
64. Stefuca V, Gemeiner P, Kurillova L, Danielsson B, Bales V (1990) Enzyme Microb Technol 12:830
65. Danielsson B, Flygare L, Velev T (1989) Anal Lett 22:1417

Thermal Biosensors in Biotechnology

Frank Lammers · Thomas Scheper

Institute for Technical Chemistry, Callinstraße 3, D-30167 Hannover, Germany
E-mail: frank.lammers@hmrag.com
E-mail: scheper@mbox.iftc.uni-hannover.de

The application of enzyme thermistor devices for the continuous monitoring of enzymatic processes is described. Different hardware concepts are presented and discussed, practical results are also given. These devices were used to analyze the enantiomeric excess in biotransformation processes and for thermal immunoanalysis. In addition, the biosensors were applied for the monitoring and control of an L-ornithine producing process and for the application in hemodialysis monitoring. A review section discusses the use of thermal biosensors for monitoring biotechnological processes in general.

Keywords: Biosensors, Enzyme thermistor, Process monitoring, Enzyme Technology, Biotransformation.

Advances in Biochemical Engineering/
Biotechnology, Vol. 64
Managing Editor: Th. Scheper

1 Introduction

Over the last two decades, great enthusiasm has been observed in biosensor research. A key position for biosensors was predicted in analytical sciences allowing a sensitive, selective and, above all, a comparatively cheap determination of nearly all interesting substances. Therefore, Prognos AG, Basel, expected world-wide sales of about 2.3 Million DM in the year 2000, and many investors promoted R & D on a large-scale. Nowadays, the enthusiastic discussion has mostly disappeared caused by disillusionment due to technical problems like biosensors long-term stability *and* economical aspects like the convincing benefits of biosensors. In most cases, a considerable need for R & D was realised in order to fulfil customers requirements.

In the last five years, electrodes and optrodes have found wide-spread use in biosensorics and seemed to be the most promising and successful transducer techniques. Electrodes were predicted as having an especially good future, and several companies have developed biosensors on the basis of electrochemical transducers (e.g., Anasyscon GmbH, Hannover, Germany; Biometra GmbH, Göttingen, Germany, and Ismatec AG, Glattbrugg, Switzerland). The main reason for the use of electrodes has been the long experience and knowledge in producing them even in miniaturized size, bulk quantity and good reproducibility. Nowadays, electrodes are comparatively cheap bulk products.

Thermal biosensors have attracted less consideration. Moreover, adverse comments like complicated thermostating, very weak sensitivity or non-specific heating effects have resulted in a poor reputation. Actually, this trend is surprising because thermal biosensors have influenced the whole of biosensor research over and over again (Mosbach, 1991). Especially the enzyme thermistor (ET) has enriched our knowledge about immobilized multi-enzyme systems for signal amplification, the use of immobilized coenzymes and different immobilization techniques. Moreover, ET basic research was decisive for immunosensorics or concanavalin-A-based reversible biosensors. Thermal biosensors have multiple advantages:

- Due to there being no chemical contact between transducer and sample, thermistors have very good long-term stability.
- Thermistors are cheap bulk products.
- Measurements are not disturbed by varying optical or ionic sample characteristics.

- In some cases, thermal biosensors work without complicated and interference-prone multi-enzyme systems, e.g., disaccharide analysis.
- Thermal biosensors have found multiple applications.

In this article, we would like to review the principles and applications of thermal biosensors. Especially, the newest results of ET research and trends in hardware developments are pointed out. Thermal biosensors will probably have a promising future in biotechnology.

2
Thermal Biosensors: Principles and State of the Art

Nearly all biochemical reactions are of exothermic character, i.e., an enzymatic conversion of a substrate is accompanied by heat production. The first law of thermodynamics decribes a proportional relationship between the heat produced and amount of molar enthalpy:

$$Q = -n_p \Sigma \Delta H$$

Due to heat production, a local temperature shift ΔT is observed that depends on the heat capacity C_s of the surrounding system:

$$\Delta T = \frac{Q}{C_s} = -\frac{n_p \Sigma \Delta H}{C_s}$$

Enzymatic reaction enthalpies are in the range of −10 to −100 kJ/mol and cause local temperature changes of a few mK. In the last twenty years, many experiments were described in literature using thermistors for measuring in the mK range. Thermistors are ceramic semiconductor resistances with high Ohm data and strong negative temperature coefficient (between −3 and −6%/K). Therefore, thermistors are called NTC-resistances (*n*egative *t*emperature *c*oefficient). The universal detection principle to link a nonspecific detection of heat with a highly specific enzymatic reaction rose in a number of different applications.

In the initial experiments, very simple devices were used. Partly, they had unfavourable response-times, a complicated thermostating, small sample frequencies and an irregular baseline. The thermistor was fixed at the tip of a flow through coil (cartridge or tube with immobilized enzyme). The enzymatic reaction takes places in the coil and is accompanied by heat production. The flowing medium transports the resulting temperature gradient to the fixed thermistor that detects the local temperature change. Therefore, the thermistor data are linked with the enzymatic conversion. Several authors have described arrangements of this kind (Mosbach and Danielsson, 1974; Cannings and Carr, 1975; Schmidt et al., 1976).

At the University of Lund in Sweden, Mosbach and Danielsson (1981) developed the first operational ET systems. These are still being supplied in small series by Thermometric Co, Järfalla, Sweden or ABT, Lund. Figure 1 schematically shows the ET setup. It consists of an external aluminium cylinder that is

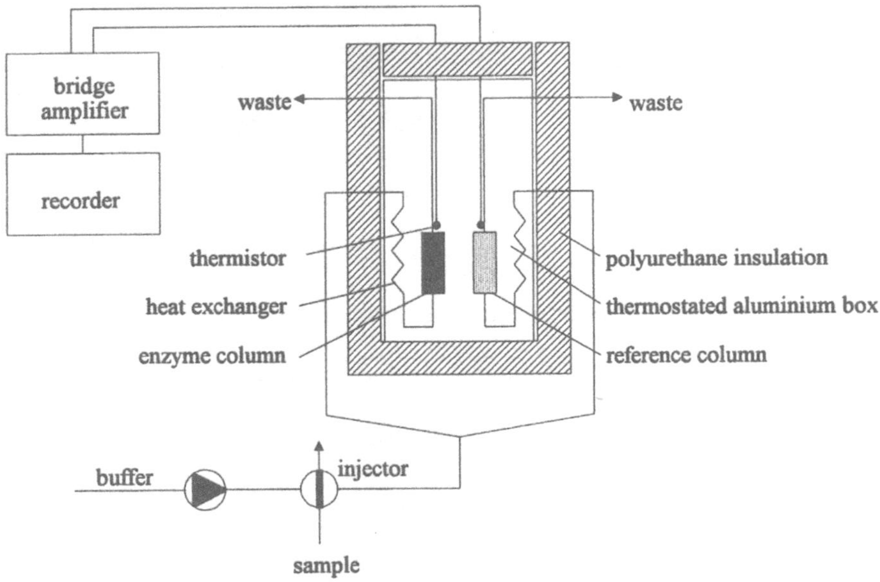

Fig. 1. Principle set-up of an enzyme thermistor

thermostated via a proportional controller at physiological temperatures (25, 30 or 37 °C ± 0.01 °C). An inner aluminium cylinder contains two fastenings for a measuring and reference thermistor. A box filled with polyurethane foam insulates the aluminum cylinders. Samples are injected via the FIA-principle and pumped to the ET. Here, the aluminium cylinder thermostats the buffer stream that flows through thin-faced steel tubes (0.8 mm inner diameter). The tubes are connected with gold capillaries (good heat exchange) with fixed thermistors (type: GB42JM65, 16 kΩ at 25 °C; Fenwal Electronics, Framingham, MA, USA) and interchangeable columns containing the immobilized enzyme. After an enzymatic conversion, the heated sample flows through the gold capillary and reduces thermistor resistance. A Wheatstone bridge registers the signal, and a chopper stabilized amplifier (MP221, Analogic Corp.) indicates a voltage. The ET registers about 80% of the heat produced. In order to minimize the effect of mixing enthalpies, a two channel version is obvious: one channel with immobilized biocatalyst, and a second reference channel with an inactive column.

Sauerbrei (1988) developed a multi channel calorimeter for the determination of up to three different analytes. A two point controller grossly thermostats the aluminium cylinder to a desired temperature, and a PI controller ensures fine tuning. A multiple bridge is connected with an amplifier and an 8-bit A/D board, and a microprocessor takes care of data acquisition and analysis (peak height and area).

Based on experiences with the Lund-ET, Hundeck et al. (1992) continued the development of multi channel ET and constructed a stand-alone version for the simultaneous determination of up to four analytes (Fig. 2). Here, thermistors

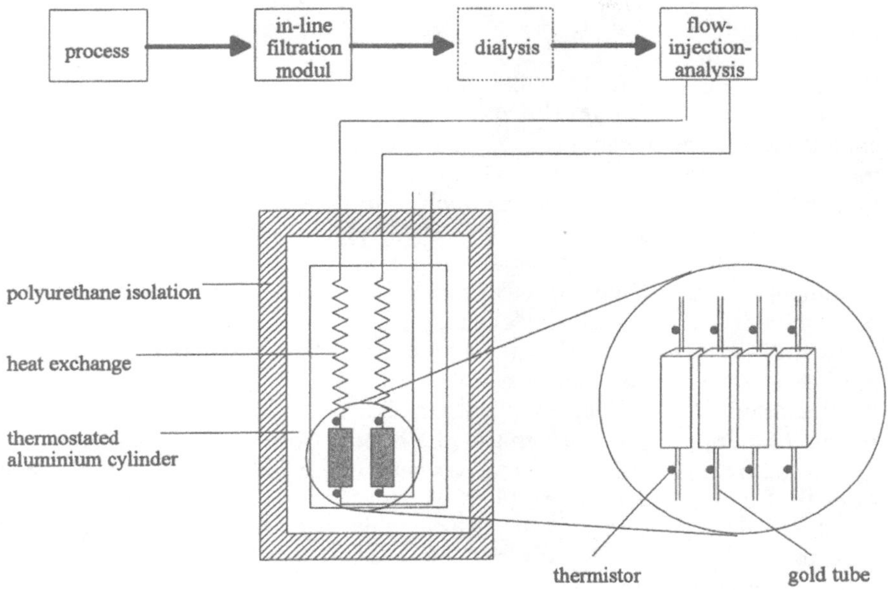

Fig. 2. Four channel ET for on-line bioprocess control

were fixed at the cartridge inlet and outlet. The difference method enables a finer measurement. The design allows a fast interchange of exhausted enzymes. A 22-bit A/D board completely covers the range of the wheatstone bridges. Therefore, manual equalizing is not necessary. Additionaly, the system allows us to acquire up to eight different sensor signals (e.g., pO_2- or pCO_2 data), and controls up to twelve valves or pumps.

The multi channel ET sucessfully managed a fully automated bioprocess monitoring and control. Hundeck et al. (1992) used it in several bioprocesses for monitoring different sugars simultaneously (glucose, maltose, sucrose and lactose). Moreover, the system was used for enantioselective monitoring of amino-acid esters, and detailed investigations were performed with immobilized microorganisms. Nevertheless, the system is very unwieldy, the operation software too complicated and the electronic modules outdated.

In the last five years, hardware development has been focused more on miniaturization and new sensor concepts. Due to customers requirements, a need for simpler operation was realized as well.

3 Applications of Enzyme Thermistors – A Review

On the basis of a universal detection principle, the ET has attracted wide attention for several analytical procedures. In this section, applications are pointed out that have been of great interest in the last decade and explains ETs' attraction in bioanalytical sciences.

3.1 Clinical Analysis

Worldwide, about 12 to 15 billion US$ per annum are spent for analytical purposes. Nearly 50 million US$ of this sum is devoted to enzymes. In clinical chemistry, several metabolites and their concentrations give important information about patients' health (see Table 1).

Due to the development of new immobilization techniques, the routine use of enzymes in clinical analysis was given a tremendous fresh impetus. Immobilized enzymes enable multiple applications, simple operation and an essential simplification of analyzers. Thus, biosensors including the ET were predicted to hold a predominant position in clinical chemistry.

Most clinical analysis deals with blood or urine metabolites in the micro- and millimolar range. Table 2 shows a collection of low molecular weight analytes that were successfully determined with the ET. So far, its use is limited to a few research laboratories. On account of the high expenditure and relatively low measuring frequency (about 12 analyses per hour), a wider acceptance has been difficult (Scheller and Schubert, 1989). Moreover, economical aspects have to be taken into account. Especially in medical analysis, biosensors have to compete with the well-established test strips. Companies like Boehringer Mannheim (Mannheim, Germany), Bayer Diagnostics (Munich, Germany) or Merck (Darmstadt, Germany) supply disposable tests with a complete enzyme chemistry including simple-to-operate pocket devices. Although these tests have only moderate precision, they fulfil a good marketing strategy due to high unit costs. Thus, disposable tests are of interest for patients to be able test themselves and for small clinical laboratories. Due to these aspects, biosensors for clinical analysis – including the ET – will probably only be applied where high sampling frequencies or on-line-analysis are of interest (e.g., bed-side monitoring).

3.2 Immunoanalysis

In biochemistry, non-enzymatic proteins are analyzed by immunochemical methods. Especially the popular enzyme-linked immunosorbent assay (ELISA)

Table 1. Important clinical analytes and normal ranges in blood (Pschyrembel, 1993)

Analyte	Diagnostical use	Range [mmol/l]
cholesterol	arteriosclerosis	2.62–7.62
creatinine	kidney diseases	0.05–0.11
glucose	diabetes mellitus	3.60–5.60
uric acid	arthritis urica (gout)	0.12–0.40
urea	kidney diseases	3.60–5.40
lactate	liver diseases	1.00–1.78
triglycerides	arteriosclerosis	0.45–1.70

Table 2. Enzyme thermistors for clinical chemistry

Analyte	Immobilized enzyme	Concentration range [mmol/l]	Reference
ascorbic acid	ascorbate oxidase	0.05 – 0.6	Mattiasson et al. 1982
ATP	apyrase	1 – 8	Mosbach/ Danielsson 1981
cholesterol	cholesterol oxidase	0.03 – 0.15	Danielsson et al. 1981 a
cholesterol ester	cholestrol oxidase + cholestol esterase	0.03 – 0.15	Danielsson et al. 1981 a
creatine	creatinase + sarcosin oxidase + catalase	0.1 – 5	Lammers 1996
creatinine	creatinine iminohydrolase	0.01 – 10	Danielsson et al. 1981 a
ethanol	alcohol oxidase + catalase	0.01 – 2	Guilbault et al. 1983
glucose	glucose oxidase + catalase	0.002 – 0.8	Schmidt et al. 1976
	hexokinase	0.5 – 25	Bowers and Carr 1976
lactate	lactate-2-monooxygenase lactate oxidase+catalase	0.01 – 1 0.005 – 2	Danielsson et al. 1981 a Danielsson 1994
oxalic acid	oxalate oxidase oxalate decarboxylase	0.005 – 0.5 0.1 – 3	Winquist et al. 1985 Danielsson et al. 1981 a
pyrophosphate	pyrophosphatase	0.1 – 20	Satoh et al. 1988
triglycerides	lipoproteine lipase	0.1 – 5	Satoh et al. 1981
urea	urease	0.01 – 500	Danielsson et al. 1988
uric acid	uricase	0.05 – 4	Danielsson et al. 1981 a

have attracted great interest. Here, antibody specificity to an antigen is used for protein analysis. Due to its time-consuming nature, bioengineers require fully automized systems for process monitoring of special proteins like monoclonal antibodies or recombinant t-PA. Automated immunoanalyzers allow a real-time process control whereas the well established ELISA kits only perform a process documentation.

On the basis of an enzyme thermistor, Mattiasson et al. (1977) developed one of the first immunosensors. Immobilized antibodies against albumin are placed in a column and set into an ET. After injection of an albumin-sample and a known amount of enzyme-labeled albumin, both are separated from the sample matrix by antibody-antigen-interaction. After injection of a substrate, the change in heat is a measure of analyte concentration. The less heat produced means that more albumin has been bound. An elution step regenerates the ELISA. Due to its thermal detection principle, the procedure is called TELISA (thermometric enzyme-linked immunosorbent assay). Figure 3 shows the principle of the TELISA procedure in its sandwich configuration.

Table 3 illustrates the various proteins that have so far been determined using a TELISA. Nevertheless, the TELISA competes with recently developed fluorescence assays. The latter are cheaper, more sensitive and faster due to not need-

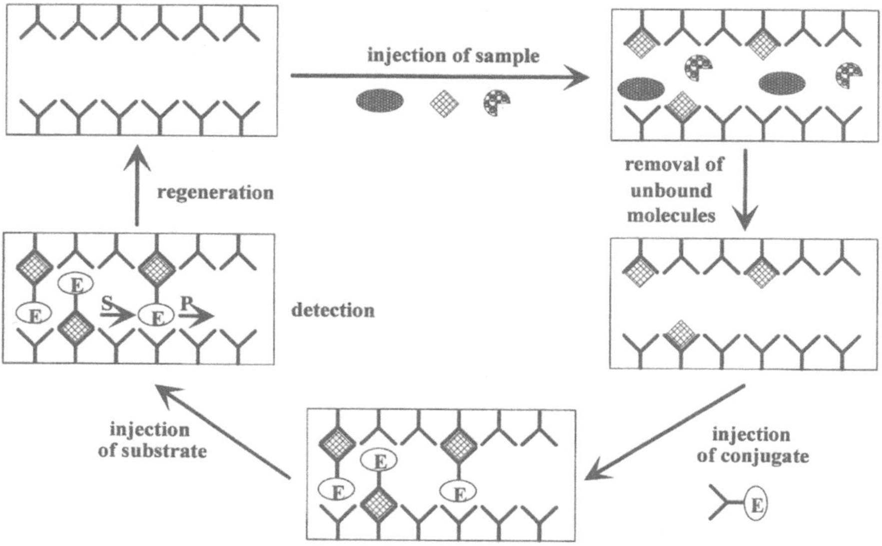

Fig. 3. Principle of a sandwich TELISA

ing of conjugates. But basic research on TELISA systems has been of tremendous importance for future immunosensors including fluorescence assays.

3.3 Determination of Enzyme Activities

In bioprocess monitoring, medicine and downstream processing, the analysis of enzyme activities is of tremendous importance. In mammalian cell cultures, lactate dehydrogenase activity informs us about the state of the cell; enzyme activities in blood, inform physicians about the patient's health; and in downstream processing, several biochromatographic procedures are monitored via enzyme activity analysis. After a simple modification, the ET can be adapted for these applications. Here, sample aliquots are mixed directly in front of the thermistor with a buffer containing substrate . Again, the resulting heat provides information about the concentration (i.e. enzyme activity). The technique can be used for enzyme activities up to 0.01 units/ml (Danielsson and Larsson, 1990). Danielsson et al. (1981b) demonstrated this procedure by monitoring chromatographic purifications of enzymes. Here, enzyme activity is measured at the efflux of a chromatographic column. Experiments were carried out with ionic-, gel-, and affinity chromatography to separate lactate dehydrogenase, glucose-6-phosphate dehydrogenase and hexokinase (see Table 4).

In another procedure, substrate or product concentration is analyzed after an enzymatic conversion has taken place. The method extends the possibilities of analysing enzyme activities. For instance, arginase cleaves buffer-containing arginine to urea and ornithine.Therefore, the ET determines arginases activity

Table 3. TELISA systems

Analyte	Immobilized compound	Concentration range [µg/ml]	Reference
albumin	immobilized antibody	> 10^{-10} mmol/l	Mattiasson et al. 1977
gentamicin	and enzyme-labeled	> 0.1	Danielsson et al. 1981a
insulin	antigen	0.1 – 50	Birnbaum et al. 1986; Mecklenburg et al. 1993
human IgG (polyclonal)	protein A	50 – 500	Lammers 1996
	protein G	50 – 600	Lammers 1996
rabbit IgG (polyclonal)	protein A	10 – 150	Lammers 1996
	protein G	20 – 1000	Lammers 1996
	anti-rabbit-IgG	10 – 200	Grau 1993
mouse IgG (monoclonal)	protein A	100 – 2000	Brandes 1992
	anti-mouse-IgG	5 – 200	Lammers 1996
rt-PA	anti-rt-PA-IgG	5 – 50	Lammers 1996

Table 4. Enzyme activities determined with an ET

Enzyme	Mixed with	Concentration range [units/ml]	Reference
acetylcholine esterase	no mixing; immobilized choline oxidase + catalase	0.2 – 4	Lammers 1996
arginase	no mixing; immobilized urease	0.39 – 100	Lammers 1996
catalase	H_2O_2	20 – 60	Sauerbrei 1988
β-galactosidase	lactose	0.5 – 4	Hundeck 1988
glucose oxidase	no mixing; immobilized catalase	0.29 – 80	Lammers 1996
glucose-6-phosphate hydrogenase	glucose-6-phosphate + NAD^+	0.2 – 3.0	Danielsson et al. 1981b
hexokinase	glucose + ATP	0.1 – 2.5	Danielsson et al. 1981b
lactate dehydrogenase	pyruvate + NADH	0 – 20	Flygare et al. 1990
peroxidase	H_2O_2	0.2 – 4	Hundeck 1988
urease	urea	0.1 – 100	Danielsson 1979

via urea release. Figure 4 shows schematically a set-up for arginase monitoring during its chromatographic FPLC purification. The computer-controlled system facilitates a fast fraction-wise allocation of arginase activities.

Flygare et al. (1990) demonstrated ET monitoring of an affinity-adsorption procedure. Lactate dehydrogenase (LDH) was recovered from a crude solution by affinity binding to a N^6–(6-aminohexyl)-AMP-sepharose gel. The LDH activity signal from the ET was used in a PID controller to regulate the addition of

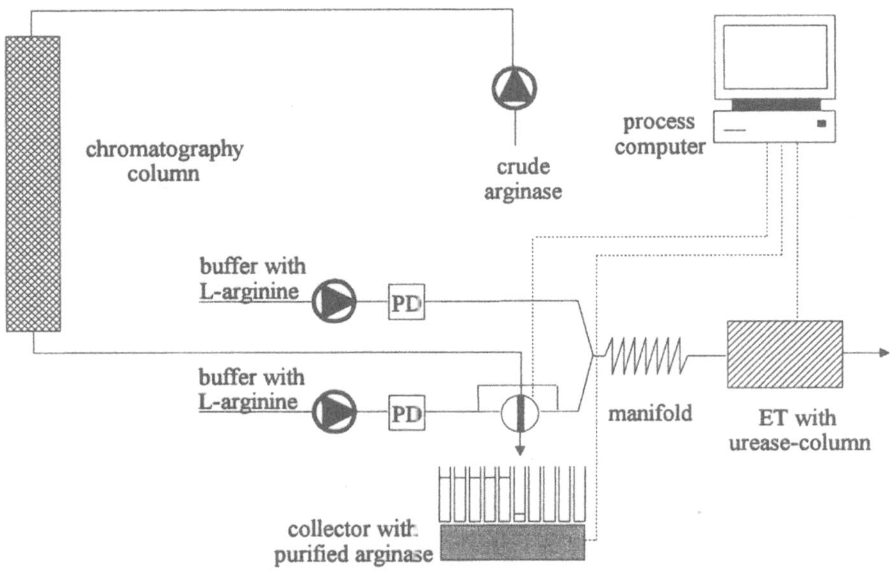

Fig. 4. Purification of arginase via FPLC and ET analysis

AMP-sepharose suspension to the LDH solution. Both examples, arginase and lactate dehydrogenase monitoring, might be attractive in industrial scale purification of enzymes.

Due to their limited stability, the use of immobilized enzymes might be problematic for monitoring enzyme production (e.g., urease for arginase monitoring). Optical methods which do not need immobilized biocompounds should be more profitable in long-term procedures. Nevertheless, special dyes or substrates with varying optical characteristics after an enzymatic conversion must have at one's disposal.

On-line-monitoring of enzyme production during industrial fermentation processes are problematic because sampling probes are still not adequate for long-term use. In particular, clogging falsifies the real enzyme activity. Moreover, a lot of enzymes are produced intracellularly and complicate a simple monitoring procedure. Therefore, this kind of monitoring is still under development.

3.4 Process Monitoring

The ET has been employed in several biotechnological processes. This section focuses on selected experiments shown in Table 5.

Mattiasson et al. (1983) used the ET for monitoring and control of a cultivation of immobilized *Saccharomyces cerevisiae*. Here, sucrose and ethanol were monitored on-line, and the data were used for computer-controlled sucrose feed.

Table 5. ET applications for process control

Analyte	Immobilized compound	Concentration [mmol/l]	Reference
acetaldehyde	aldehyde dehydrogenase	not mentioned	Rank et al. 1995
cellobiose	*β*-glucosidase + glucose oxidase + catalase	0.05–5	Danielsson et al. 1981a
cephalosporin	cephalosporinase	0.005–10	Danielsson et al. 1981a
ethanol	alcohol oxidase + catalase	0.002–1	Rank et al. 1995
galactose	galactose oxidase	0.01–1	Mattiasson et al. 1982
L-glutamine	glutaminase + glutamate decarboxylase	0.5–50	Lammers 1996
glycerol	glycerokinase	not mentioned	Rank et al. 1995
L-lactate	lactate oxidase + catalase	0.05–2	Danielsson et al. 1994
lactose	lactase + glucose oxidase + catalase	0.05–10	Mattiasson et al. 1982
maltose	*α*-glucosidase + glucose oxidase	0.15–6	Hundeck et al. 1992
monoclonal antibodies	anti-mouse-IgG	5–200 µg/ml	Lammers 1996
penicillin G	*β*-lactamase	0.05–500	Mattiasson et al. 1981
penicillin V	penicillin acylase	0.5–150	Rank et al. 1993
proinsulin	anti-proinsulin-IgG	0.1–50 µg/ml	Birnbaum et al. 1986
sucrose	invertase	0.1–100	Hundeck et al. 1992

Hundeck et al. (1992) employed a four channel ET for on-line-monitoring of glucose, maltose and sucrose during cultivation of *Bacillus licheniformis*. In this process, microorganisms produce subtilisin, a serine protease added to washing powders to improve washing efficiency. In fed-batch cultivations of *Bacillus licheniformis*, sucrose feed is a very important parameter because sucrose concentration correlates with subtilisin production. Optimal enzyme production is achieved only within a defined sucrose range. Therefore, ET sucrose data were used for maintaining a defined concentration range.

Figure 5 shows on-line-data of sucrose concentrations and the enzyme production rate obtained during a fed-batch cultivation. Obviously, sucrose concentration and enzyme activity correlate. Due to growth of the microorganisms, sucrose concentration decreases in the first five hours. After a lag, subtilisin is produced at an optimum rate. The enzyme activity decreases after that because the sucrose concentration is too low of a limited. Thus, sucrose is fed to the medium and subtilisin production increases again.

In addition, the response time of immobilized yeast to different sugars was used for monitoring and control of *Bacillus licheniformis* cultures (Hundeck et al. 1992). Specificity is not a characteristic of cell-based sensors; rather, they are characterized by their ability to react in an integrated way to various stimuli.

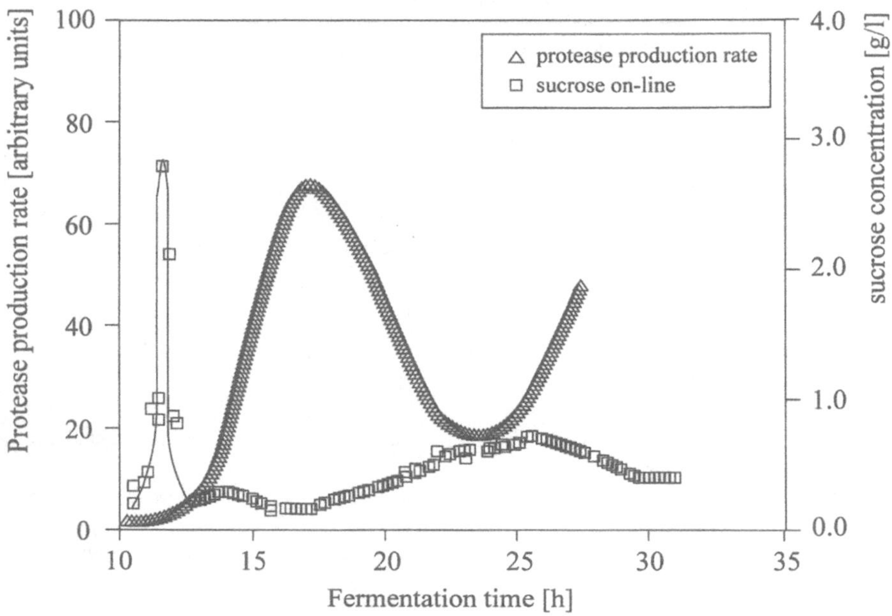

Fig. 5. Fed-batch cultivation of *Bacillus licheniformis*

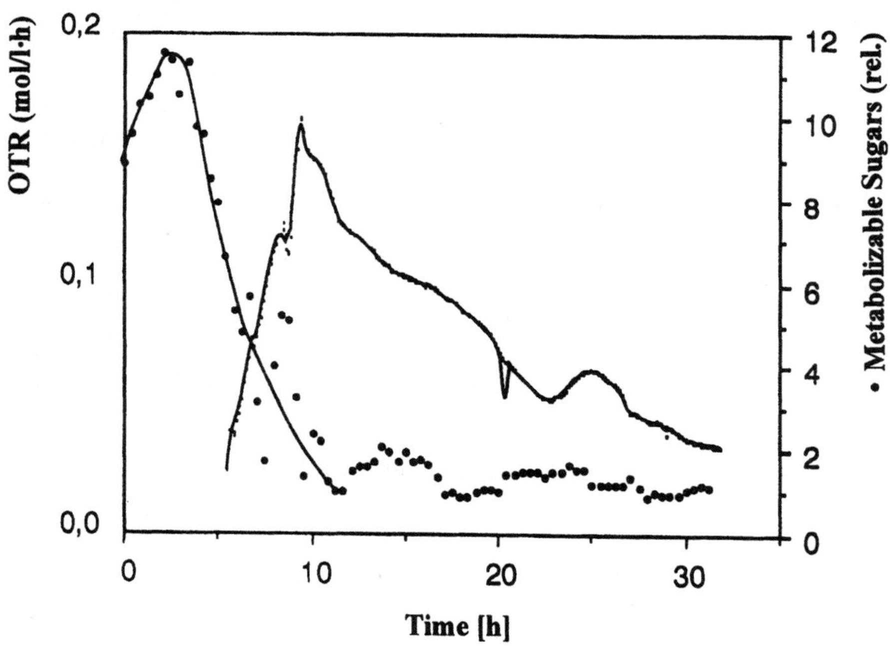

Fig. 6. *Bacillus licheniformis* cultivation. The ET with immobilized *Saccharomyces cerevisiae* was used for monitoring metabolizable sugars

This property allows us to monitor some parameters such as metabolizable sugars. Obviously, the parameter closely correlates with the oxygen transfer rate (OTR; Fig. 6). Sugars are metabolized with consumption of oxygen. Thus, the OTR increases with decreasing sugar concentrations.

Rank et al. (1992) used an ET in several industrial fermentation processes at Novo Nordisk, Denmark. The production of penicillin V in a 160-m^3 bioreactor was monitored on-line using an enzyme column with penicillin V acylase. In comparison to β-lactamase, the enzyme has a superior substrate specificity. Off-line samples were analyzed via liquid chromatography and correlated very well with on-line ET data (Fig. 7).

Several 500 l fed batch cultivations with *Saccharomyces cerevisiae* were monitored in the pilot-plant at Novo Nordisk as well (Rank et al. 1995). Here, ethanol was monitored using coimmobilized alcohol oxidase and catalase, acetaldehyde by aldehyde dehydrogenase and glycerol by glycerokinase. Acetaldehyde and glycerol analysis had to be performed with the cosubstrates β-NAD and ATP, respectively. Therefore, the monitoring was comparatively expensive. Nevertheless, ET data gave an interesting insight into yeast behavior during an industrial cultivation process.

Industrial plants are characterized by high humidity, ambient temperatures up to 40 °C, large temperature variations, water and steam outlets, and continuous vibration. The complex composition of fermentation broth and high and variable concentrations of certain components are additional problems.

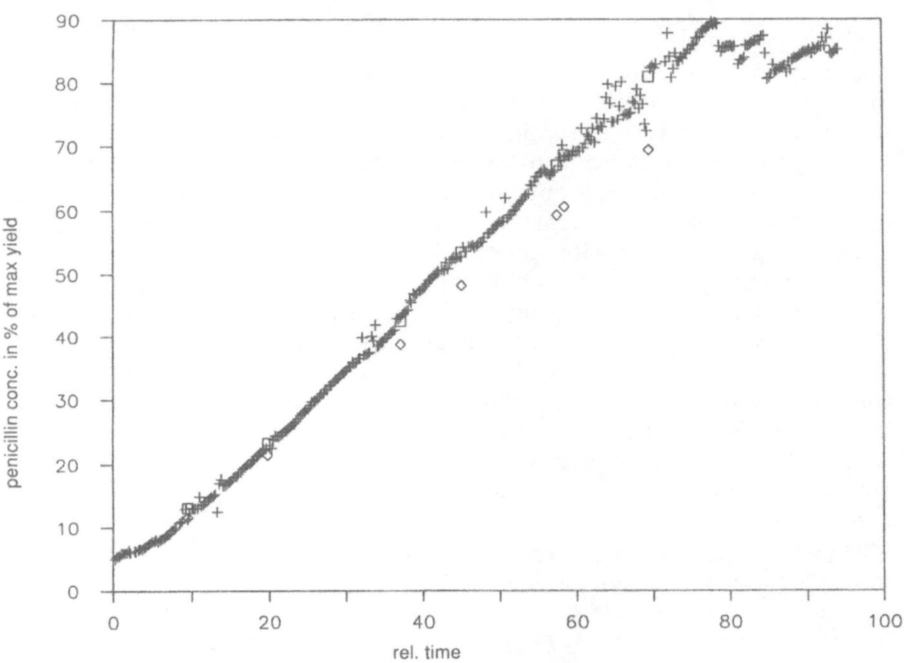

Fig. 7. On-line monitoring of a penicillin V fermentation (Rank)

Therefore, Rank et al. (1995) constructed an automated ET for process monitoring. The equipment was installed inside a steel cabinet with cool, filtered air to keep the temperature sufficiently constant. The enzyme column had to be protected against microbial growth by adding sodium azide to the buffer solution. These facts show an obvious strong difference between biosensor employment in research laboratories and industrial plants.

TELISA systems for on-line-monitoring of proteins have not been applied so far. Nevertheless, several TELISAs have been applied for off-line analysis of cultivation samples. Birnbaum et al. (1986) used a competitive TELISA for human proinsulin analysis produced from recombinant *Escherichia coli*. Lammers (1996) describes a sandwich TELISA for monitoring monoclonal mouse antibodies. Here, real samples of a hybridoma cell cultivation (Dr. Karl Thomae, Germany) were analyzed and correlated well with conventional ELISA data. A whole measuring cycle including incubation, detection and reequilibration needs 18 minutes. Due to the long cultivation time (two weeks and more) for mammalian cells, it would be sufficient for on-line monitoring. But TELISAs have other decisive disadvantages. Enzyme-labeled antibodies or antigens are expensive and need a complicated monitoring performance. Thus, procedures working without conjugates like fluorescence immunoanalysis might be more attractive.

In conclusion, ET employment is possible for on-line monitoring of industrial fermentation processes under real conditions. Nevertheless, each process has its own characteristics and needs a special monitoring concept. In special cases, sensors other than the ET might be better for monitoring and control.

3.5 Environmental Analysis

Microorganisms are universal sensors for pollutant analysis. Toxins influence the microorganisms' metabolism and are recognized via monitoring the state of the cells. In comparison to specific biosensors, the unspecific response of the microorganisms to different substances is a crucial advantage for toxicological studies. Nevertheless, simple operation with immobilized microorganisms is of tremendous importance for employing these kinds of toxiticy tests. Microbial sensors might be attractive for domestic and industrial waste-water monitoring.

Sudden changes in metabolism bring about a fast response in temperature. Thus, the ET has good prerequisites for use as a toxin guard system. Hundeck (1991) investigated the response behavior of immobilized *Saccharomyces cerevisiae* to different inhibitors (azide, Hg^{2+}, cyanide, arsenate, 2,4-dinitrophenol). Here, inhibitors were mixed with glucose, and after substrate injection, the combined substrate-inhibitor-solution was used. Increasing inhibitor concentration caused decreasing temperature signals. The inhibitors azide, cyanide and arsenate allowed multiple determinations to be made whereas Hg^{2+} irreversibly damage the cells. The substance 2,4-dinitrophenol (2,4-DNP) increases the temperature signals by decoupling the respiration chain (i.e. oxidation of NADH and decoupling of ATP production).

Some different examples showing relevance to environmental monitoring are listed in Table 6. In addition to microbial systems, immobilized enzymes that are

Table 6. ET as toxing guard system

Analyte	Immobilized compound	Concentration [mmol/l]	Reference
amines	monoamine oxidase	0.001 – 0.7	Svensson et al. 1979
azide	*Saccharomyces cerevisiae*	10 – 100 mg/l	Hundeck 1991
Cu^{2+}	urease	0.1	Mattiasson et al. 1977b
cyanide	rhodanase + injectase	0.02 – 1	Mattiasson et al. 1977a
	Saccharomyces cerevisiae	10 – 100 mg/l	Hundeck 1991
	peroxidase	20 – 100 ppm	Fischer 1989
2,4-DNP	*Saccharomyces cerevisiae*	1 – 20 mg/l	Hundeck 1991
Hg^{2+}	urease	0.01	Mattiasson et al. 1977b
	Saccharomyces cerevisiae	10 – 100 mg/l	Hundeck 1991
insecticides	acetyl choline esterase	< 0.0034	Mattiasson et al. 1979
organophosphate	hydrolytic enzymes	> 0.03	Mattiasson et al. 1979
sulfide	peroxidase	10 – 1000 ppm	Fischer 1989
Zn^{2+}	alkaline phospatase	0.01 – 1	Satoh 1991a

inhibited by special analytes are shown. Due to reversibility problems, inhibition of enzymes is a difficult procedure (Schmidt et al. 1995). Thus, microorganisms would be a better choice.

3.6 Enzymatic Amplification

Efficient working of ET depends on enzymatic reactions accompanied by high reaction enthalpies. A reaction with low enthalpy is often amplified by a sequential or cycling system. In a sequential system, a product of the first or subsequent reaction is converted by a coimmobilized enzyme with a high enthalpy effect. For instance, several oxidases produce hydrogen peroxide. Due to low reaction enthalpies of oxidases, coimmobilized catalase effects an amplification in a sequential step ($\Delta H = -100.7$ kJ/mol). Another example is arginase analysis. Here, an enzyme column with coimmobilized arginase and urease is employed. Arginase cleaves arginine to ornithine and urea whereas urease converts the latter with a high enthalpy effect to ammonia and carbon dioxide. A column without urease gives much lower thermometric signals.

The enzymatic cycling systems are characterized by a combination of enzymes (mostly oxidases and dehydrogenases) that are capable of a multiple regeneration of a substrate. Here, a repeated regeneration of the substrate causes an accumulation of heat and amplifies the signals. Enzymatic cycling systems have an important requisite, namely the analyte concentration has to be well below its Michaelis-Menten-value K_M. Otherwise, the reaction speed is not proportional to the analyte concentration (Bergmeyer, 1983).

The combination of cycling systems gives very high amplification factors. The latter describes the ratio of amplified and non-amplified signal in the linear

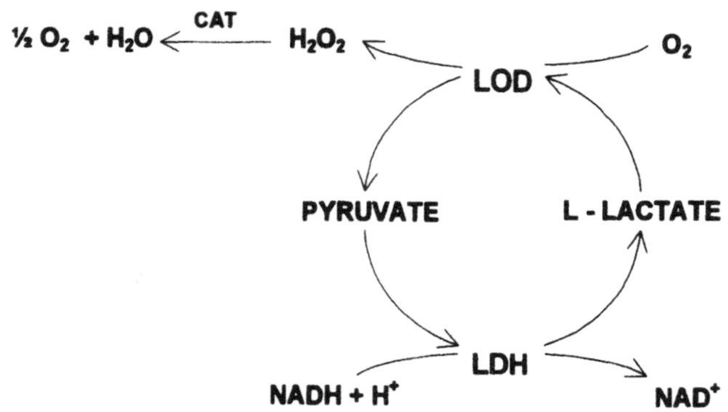

Fig. 8. Amplification system of lactate oxidase (LOD), lactate dehydrogenase (LDH) and catalase (CAT)

range. Kirstein et al. (1989) describe an ET analyzing ATP with a detection limit of 10 nmol/l. The enzymatic cycling system causes an amplifiction factor about 1700. Another amplification system uses a combination of lactate oxidase, lactate dehydrogenase and catalase (LOD/LDH/CAT-system; Fig. 8). Due to evolution of hydrogen peroxide from the cycling reaction, a coimmobilisation of catalase effects an additional amplification.

The cycle represents an oxidation of NADH with oxygen accompanied by a very high enthalpy ($\Delta H = -255$ kJ/mol). The value agrees with the sum of enthalpies taking part in the cycle (Scheller et al. 1985). Mecklenburg et al. (1993) used the LOD/LDH/CAT-system for an insulin-TELISA. Here, an amplification factor of about 10 was observed. Nevertheless, the cycling system is expensive, complicated and difficult to reproduce (Lammers, 1996). Especially for TELISA procedures, an optimized substrate for peroxidase labeled antibodies was developed. The substrate (2 mmol/l H_2O_2 and 2 mmol/l aminoantipyrine) causes a similar TELISA sensivity to that of the LOD/LDH/CAT cycle but is much easier to use, cheaper and very reproducible (Lammers, 1996).

4
New Fields for Enzyme Thermistors

4.1
Enantiomeric Analysis

In the pharmaceutical industry, the production of enantiomerically pure substances is of tremendous importance because the same compounds but of opposite chirality often have extremly different therapeutic effects. Thus, enantiomeric analysis is of special importance. Nowadays, enantioselective syntheses are monitored via time-consuming, expensive and complicated procedures (e.g., chiral GC and HPLC). Biosensors might be an interesting alternative or a supplement to existing techniques. Here, immobilized enzymes with different enantiospecifities can be used to analyze the ratio of enantiomers. Figure 9 shows the principle of two different enantioselective biosensors. In the first method, one sensor specifically detects a D-enantiomer whereas the second one analyzes the concentration of a L-enantiomer. Method 2 shows a combination of an unspecific and a specific working sensor. Here, a comparison of both sensor signals allows a fast determination of the ratio of enantiomers.

Hundeck et al. (1993) showed the ET analyzing racemates of DL-phenylalanine-methyl ester. Here, unspecific pig liver esterase cleaves both antipodes whereas α-chymotrypsin converts only the L-enantiomer (Fig. 10). Both reactions release protons that are easily detected by tris/HCl-buffer (enthalpy of protonation: $\Delta H = -47$ kJ/mol).

Lammers (1996) extended the esterase/α-chymotrypsin system with DL-tryptophan-, DL-tyrosine-, DL-methionine- and DL-phenylglycinemethyl ester. All substances give very good signals with immobilized esterase and tris/HCl-buffer. Immobilized α-chymotrypsin converts L-tryptophan- and L-tyrosine-methylester with very good signals whereas L-methionine- and L-phenylglycine-methylester effect only a few or give no heat. In contrast, the L-phenylglycine-methylester seems to be too small for α-chymotrypsin, and L-methionine-methylester is too hydrophilic due to its side chain (Lammers, 1996).

4.2
Aminoacid Analysis

Aminoacid monitoring is of tremendous importance in biotechnology. In this section, three different ET for L-arginine, L-asparagine and L-glutamine are presented. The aminoacid L-arginine is required for an optimal growth and preservation of nitrogen equilibrium. Moreover, L-arginine is important in hormone producing processes (Alonso et al. 1995). Thus, media for mammalian cell cultivation such as Coon's F12 containing the aminoacids in relatively high con-

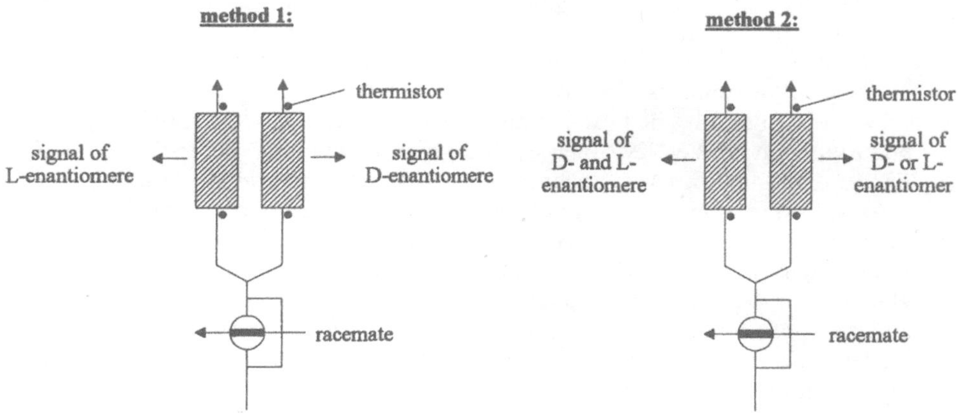

Fig. 9. Principles of enantiomeric analysis

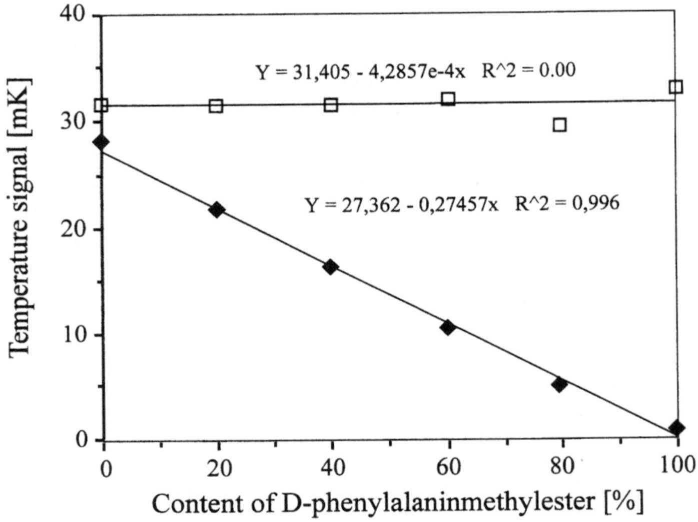

Fig. 10. Enantiomeric analysis of DL-phenylalaninemethylester

centrations (2 mmol/l and higher). The analysis of L-arginine is performed using coimmobilized arginase (1000 units) and urease (250 units) in 0.1 M potassium phosphate buffer pH 9.5. Due to the instability of urease, the buffer contains 2 mm L-cysteine as well. Arginase cleaves the aminoacid to L-ornithine and urea. Consequently, urease converts the urea to ammonia and carbon dioxide. The sequential amplification reaction showed a broad detection range (0.1 – 100 mmol/l) similar to immobilized urease columns. This might be explained by similar Michaelis-Menten-values (arginase: K_M = 11.6 mmol/l; urease: K_M = 10.5 mmol/l). The amplification factor is about fourteen.

The aminoacid L-asparagine is used as a nitrogen source in different cultivation processes. Especially, the aminoacid L-asparagine has found application in fermentation processes of ergotamine-producing *Claviceps purpurea*. Here, medium concentrations up to 10 g/l are used (Amici et al. 1967). The ET is set up for L-asparagine analysis by columns of immobilized aparagine (200 units). Asparagine deaminates the aminoacid and give good signals in the range 0.5 and 100 mmol/l. The best signals were obtained in 0.1 M potassium phosphate buffer at pH 8.6. Nevertheless, the system works in 0.05 M Tris/HCl as well. The latter might be interesting for samples with high magnesium content. In this case, potassium phosphate buffer is not suitable because magnesium phosphate causes problems in FIA-systems.

Especially in mammalian cell cultivations, the aminoacid L-glutamine and glucose represent the most important energy sources. The aminoacid is carbon and nitrogen source for the syntheses of aminoacids, fats, proteins, antibodies, nucleotides, purines and pyrimidines (Jeong and Wang, 1995). Thus, monitoring of L-glutamine is a desirable task. An ET employment is possible with coimmobilized glutaminase (50 units) and glutamate decarboxylase (100 units). Glutaminase deaminates L-glutamine with release of L-glutamate. The latter inhibits the deaminating enzyme and reduces its maximum production. Thus, glutamate decarboxylase is coimmobilized in order to remove the inhibitor. The biosensor response substantially depends on the buffer's pH (Fig. 11). Even at 100 mmol/l L-glutamine, no signal is observed with a buffer of pH 7.0. On the other hand, acid buffer (pH 4.9) cause very good signals. Thus, cultivation samples have to be imperatively acidified before analysis. This is automatically mastered in a computer controlled FIA system (Lammers, 1996).

4.3 Medical Monitoring

Due to requirements for individual medical treatment, on-line medical or bedside monitoring is one of the most promising tasks for biosensors. Especially, in hemodialysis treatment. Here, patient's blood is purified by removing enriched urea using a dialysis cell. The procedure requires about four to seven hours. A shorter hemodialysis time might be harmful whereas a long purification restricts the patient's quality of life. Thus, an individual monitoring of the purification progress is very desirable.

In hemodialysis devices, urea analysis makes it possible to monitor the effect of the medical treatment continuously. For this purpose, the ET feasibility in urea monitoring was investigated. In comparison to other transducers, the ET

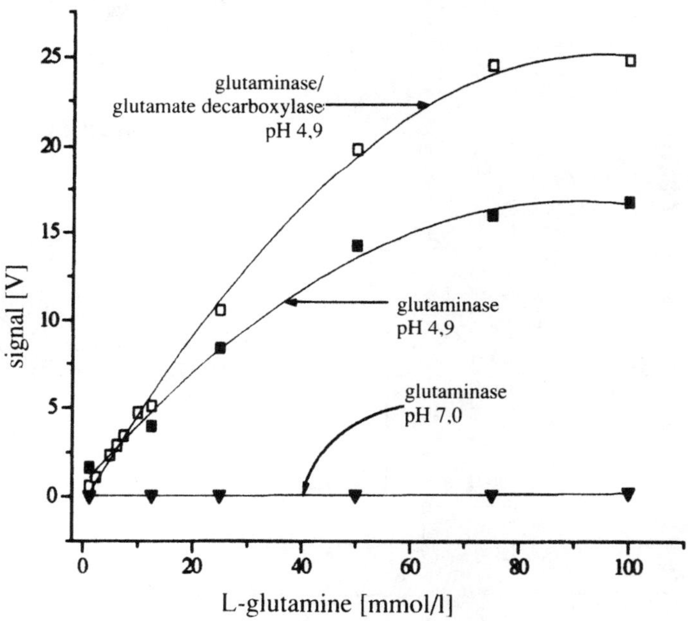

Fig. 11. Analysis of L-glutamine at pH 7.0 and pH 4.9

has a tremendous advantage in that there is no electrical contact between patient and analyzer. In the case of enzyme exhaustion, the column is removed and a fresh one installed. Thus, urease columns are cheap throw-away articles. Urea analysis is performed in the dialysis buffer and not in the patient's blood. Therefore, sterile conditions are not necessary.

Figure 12 shows a possible set-up for hemodialysis monitoring. Patients blood is pumped through a dialysis cell, and low molecular weight substances including urea are removed by a semipermeable membrane (cut off: 10 kD) and dialysis buffer. The urea enriched dialysate passes through an injection valve and enters a waste container. Due to switching the valve, a defined sample volume is pumped to the ET. Here, enzymatic conversion takes place via immobilized urease and provides information about the current urea concentration. Thus, the hemodialysis effect is automatically monitored via urea analysis and makes an individual treatment possible.

The set-up of a hemodialysis simulation is shown in Figure 13 (Lammers, 1996). Here, typical changes of urea concentrations in the dialysis buffer were monitored. Within 5.2 h a urea standard of 12 mmol/l was diluted to 2 mmol/l. The dilution process was monitored by a computer-controlled FIA system including an ET. Figure 14 shows the result of a simulation experiment. Obviously the ET is suitable for recording the whole concentration range. Danielsson (1995) observed similar curve shapes in measurements that were performed in real-

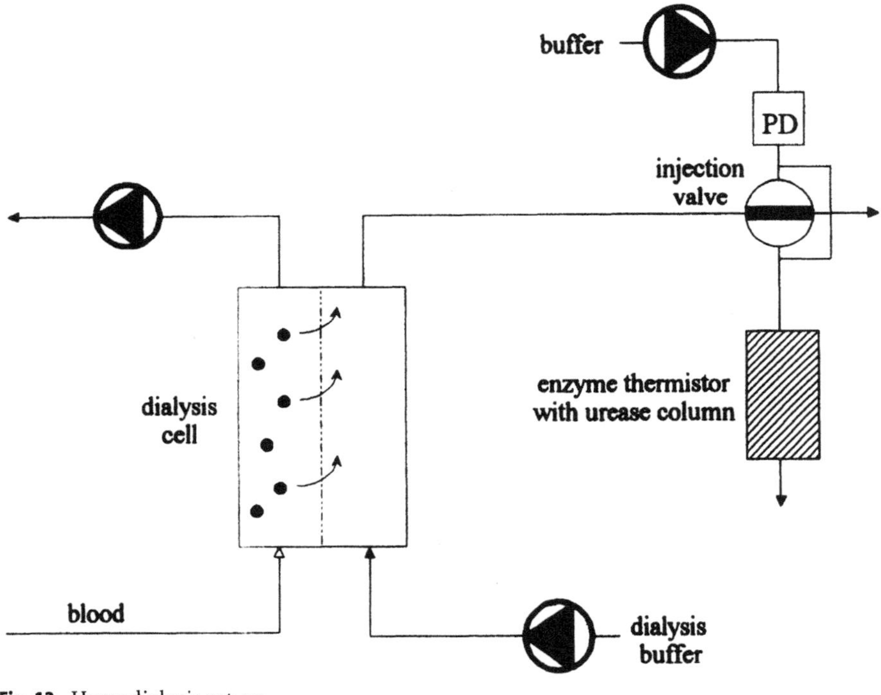

Fig. 12. Hemodialysis set-up

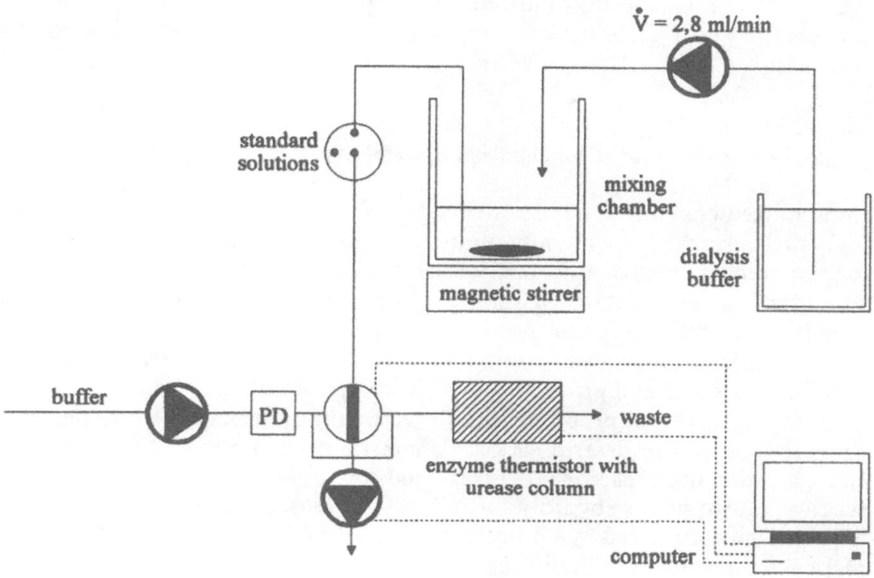

Fig. 13. Set-up of a Hemodialysis simulation

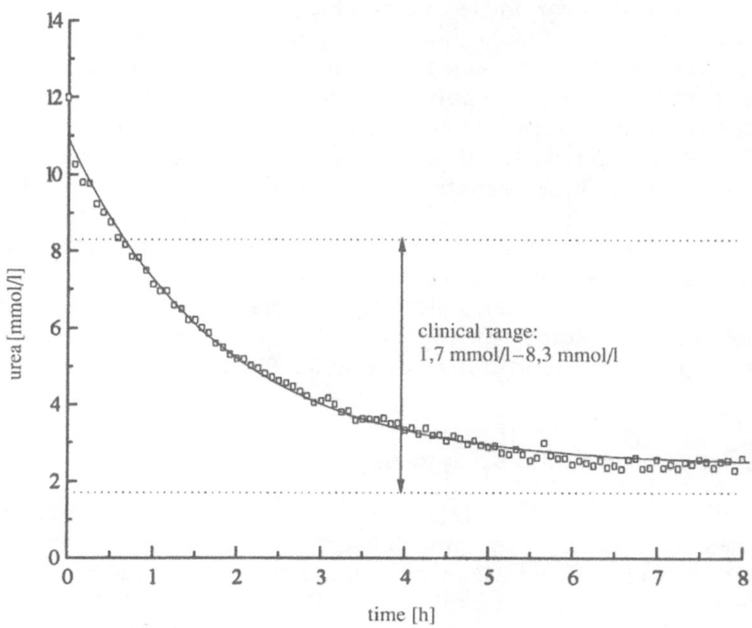

Fig. 14. On-line-data of a hemodialysis experiment

time measurements at hospitals. Moreover, patients individual urea level and ureases long term stability encourages one to employ the method in routine treatment.

4.4 Kinetic Characterization of Immobilized Biocatalysts

Immobilized enzymes are not restricted to bioanalytical applications. Increasingly they attract a huge amount of interest in industrial organic chemistry due to their excellent stereo- and enantioselectivity. Moreover, they work under mild conditions of temperature, pH and pressure. Therefore, the determination of kinetic constants is of great interest. They allow quantitative characterization of immobilized biocatalyst preparations and facilitate comparisons between different materials and procedures for biocatalyst immobilization.

Stefuca et al. (1990) proposed an ET method offering a rapid, convenient, and general approach to determine kinetic constants of immobilized biocatalysts. Here, a differential reactor (DR) was used for the measurement of the initial reaction rate of sucrose hydrolysis (Vallat et al. 1986). The enzyme column of the ET has been considered as a differential packed-bed reactor, and with a mathematical model, intrinsic kinetic constants of immobilized invertase were calculated from experimental DR and ET data.

Recently, the method was used to estimate kinetic constants of immobilized invertase in a column with different conjugates of concanavalin A (Docolomansky et al. 1994). The immobilization is based on a strong biospecific glycoprotein-lectin interaction. Due to its reversibility, this immobilization procedure is of huge interest in enzyme technology. Exhausted glycoenzymes are easy to exchange by fresh biocatalysts (Saleemuddin and Husain, 1991).

Gemeiner et al. (1993) presented a similar method for the direct determination of catalytic properties of immobilized cells. Cephalosporin C transforming *Trigonopsis variabilis* were immobilized by three different methods, filled into a column and set into the ET. After thermal equilibration, Cephalosporin C solutions (0.1 – 50 mmol/l) were continously pumped through the ET until steady-state heat production was obtained. Again, the ET was shown to be suitable for a rapid and simple estimation of the kinetic properties of immobilized cells. Microkinetic factors such as mass transfer were taken into account (Stefuca et al. 1994). Thus, ET measurements allow us to obtain intrinsic data, even from immobilized cells. Moreover, the data can be applied to optimize biocatalyst design and bioreactor models (Gemeiner et al. 1996).

4.5 Monitoring of Enzyme Catalyzed Syntheses

During the last decade, the role of enzyme catalysis in organic chemistry has increased tremendously. The availability of a huge variety of enzymes, the willingness of chemists to use them even in non-aqueous reaction phases and progress in immobilization techniques and bioreactor design led to a number of applications in bioorganic chemistry. Biosensors might be an interesting monitoring tool for enzyme catalyzed processes. The common criticism that biosensors are not stable enough for on-line monitoring, is not convincing in the field of biotransformations, because the enzymes for synthesis also have a limited process stability. On the contrary, the non-complex biotransformation media should increase their process stability.

Recently, the ET was presented as an on-line monitor for biotransformation processes (Lammers and Scheper, 1996). Here, three different enzyme-catalyzed processes of industrial interest were investigated. In the first example, the enzymatic production of L-ornithine was monitored via urea analysis. In this process, arginase hydrolyzes L-arginine with the release of urea and L-ornithine. Immobilized urease was set into the ET in order to monitor urea release, and on-line data informed us of the progress of production. Moreover, a computer-controlled set-up (Fig. 15) allowed us to remove the product at a nominal value and to add fresh substrate (Fig. 16).

In further experiments, the ET was shown to monitor enzymatic L-methionine synthesis via amino acylase. Here, decreasing starting material data (*N*-acetyl-DL-methionine) were of interest. The process was compared with native and immobilized biocatalyst (Fig. 17).

In a third example, the production of fructose was obtained. Here, the starting material sucrose was hydrolyzed via invertase to fructose and glucose. The progress in hydolysis was monitored with an ET containing coimmobilized glucose oxidase and catalase. Following the addition of glucose isomerase, it caused a shift of ingredients to fructose (Fig. 18). In summary, the connection of enzyme and (calorimetric) biosensor technology might be very effective for a number of enzyme catalyzed processes.

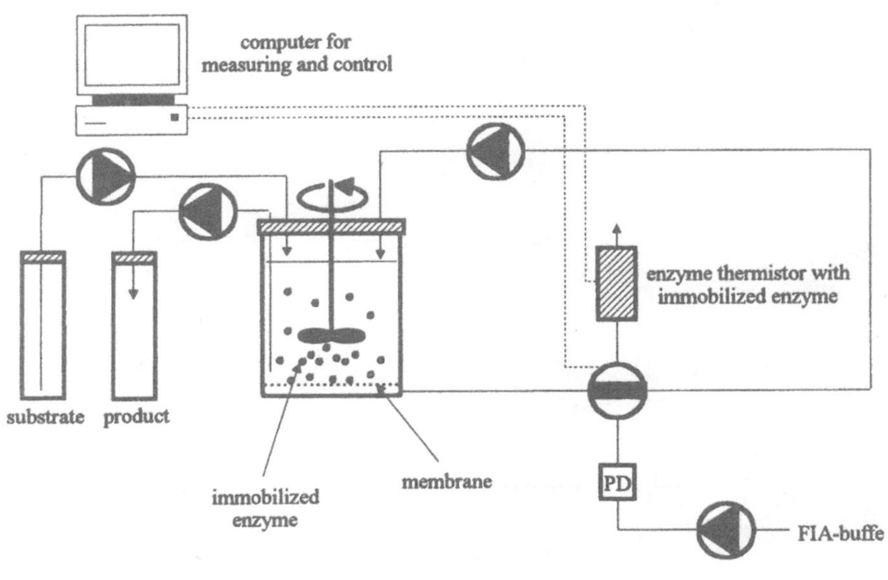

Fig. 15. Computer controlled set-up for on-line monitoring of enzyme catalyzed processes

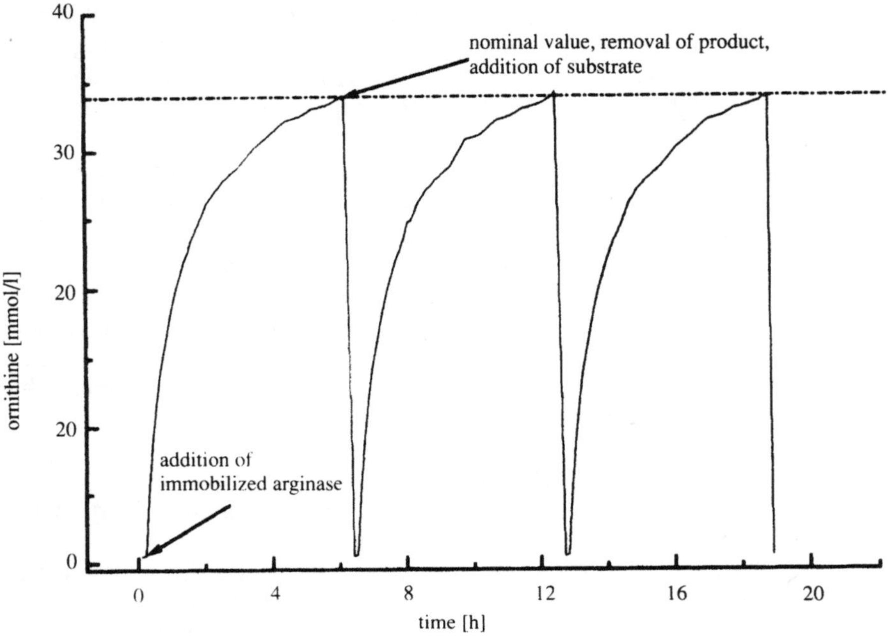

Fig. 16. Enzymatic production of L-ornithine

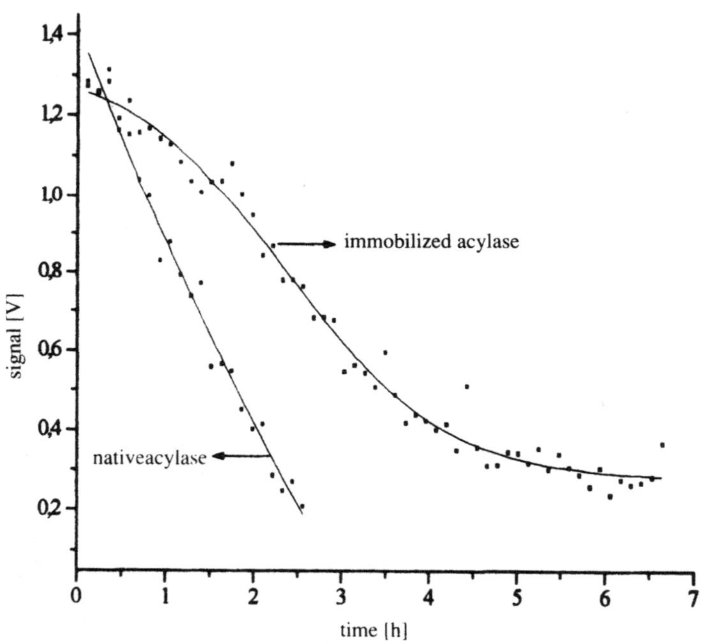

Fig. 17. Production of L-methionine via native or immobilized amino acylase

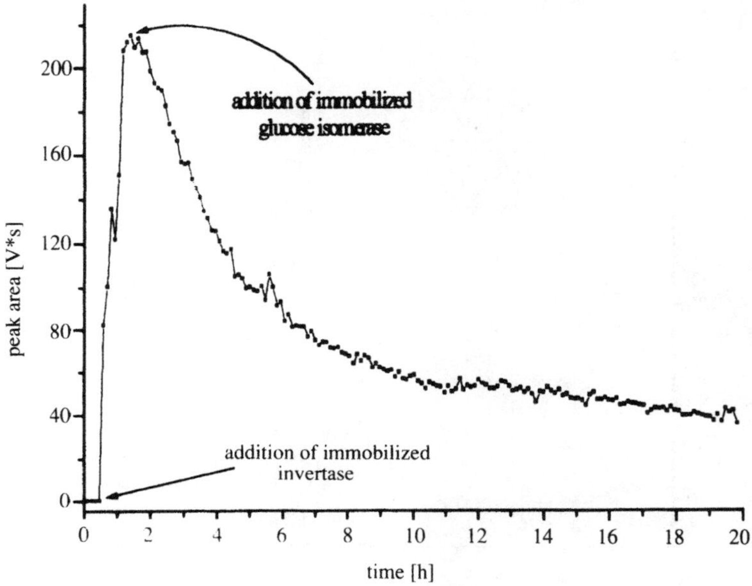

Fig. 18. Enzymatic hydolysis of sucrose following isomerization of glucose to fructose

4.6 Monitoring in Food Technology

In several European countries and the USA, the sterilization of milk is performed by addition of 0.1 % H_2O_2 as a preservative. Catalase is an ingredient of milk and decomposes H_2O_2. However, Novo Nordisk, Baegsvard, supplies catalase especially for this procedure, and Boehringer Mannheim GmbH, Mannheim, developed sensors to control remains of H_2O_2. The treatment of milk with H_2O_2 prevents "cooking flavour" which is often obtained after pasteurisation processes. Recently, Akertek and Tarhan (1995) showed that H_2O_2-sterilization has no negative effect on nutrients such as amino acids, proteins, vitamins, sugars or fats.

The ET is suitable for on-line monitoring of the raw milk sterilization process. Here, immobilized catalase is inserted into the ET, raw milk samples with H_2O_2 are automatically diluted within a FIA-system and continuously analyzed. Figure 19 shows on-line data of a H_2O_2 decomposition experiment. After addition of 0.1% H_2O_2 to raw milk, the ET signal increases tremendously whereas addition of catalase obviously causes H_2O_2 decomposition. Thus, raw milk's own catalase seems to have a too low activity.

Aspartame monitoring represents a second interesting task for the ET. Immobilized α-chymotrypsin hydrolyzes the artificial sweetener under release of protons which are easily detected via a tris/HCl-buffer due to its high protonation enthalpy. The immobilized enzyme unfortunately causes an unspecific conversion. However, the procedure might be interesting for monitoring aspartame during its industrial production within the Tosoh-process.

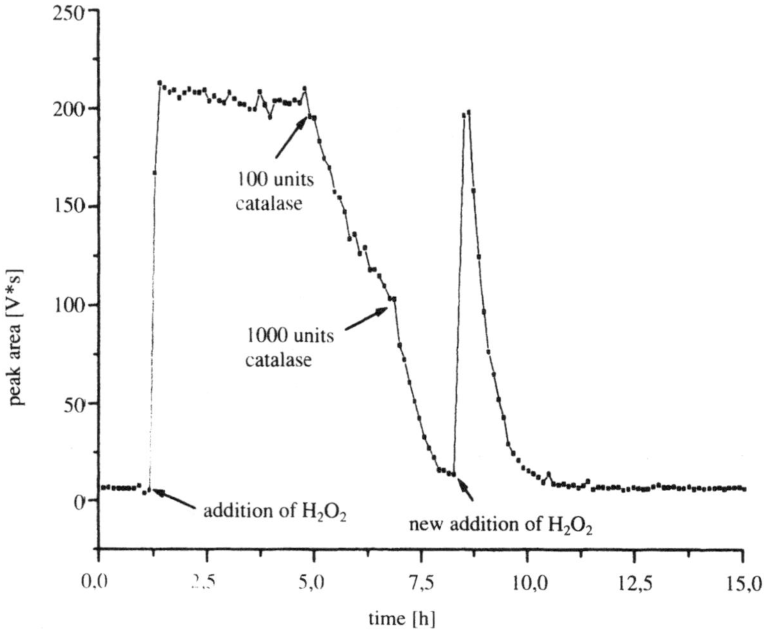

Fig. 19. On-line monitoring of milk sterilization via H_2O_2

5 New Hardware Concepts

Biosensors for clinical analysis will probably have the most economic value. Simple operating systems are required for personal home-monitoring and for small medical laboratories instead of time-consuming and expensive manual assays. The most decisive precconditions for this special kind of biosensor are miniaturized production methods, ease of operation and the possibility of using small sample volumes. Moreover, the analyzers must have convincing benefits to compete with existing dip-stick sensors. Thus, the development of microbiosensors is an active field of research. During the 1990s several researchers have investigated new methods to miniaturize thermal biosensors. Especially in basic research, new microsystem technologies have been of great interest. In this section, we would like to review these investigations and represent current trends in thermal biosensor development.

5.1 High Resolution Thin-Film Thermistors

As previously shown, thermistors exhibit high sensitivities for monitoring enzymatic conversions. Nevertheless, matched thermistor pairs for microbiosensors are not readily available from industrial suppliers which is limiting for their industrial application. Thus, Urban et al. (1991) developed a new thermal biosensor using thin-film thermistor arrays and immobilized enzymes. The miniaturized thermistor arrays were produced on glass substrates and exhibit a temperature dependence of conductivity of 2%/K, a temperature resolution of 0.1 mK and a response-time of 3 ms. The experimental set-up for glucose monitoring consists of a 3 × 2.5 × 5-cm Plexiglas block containing a flow through channel, a column with immobilized glucose oxidase and two thermistor arrays, inserted in a Peltier-thermostated aluminium block. The whole device has dimensions of 25 × 25 × 25 cm, and the thermostat has a stability of 1 mK. Thus, miniaturization has to be improved. Nevertheless, a very interesting point of this study is the high reproducibility of the thermistor characteristics due to the high level of development of thin-film technology.

5.2 Miniaturized Enzyme Thermistors

Xie et al. (1992) have developed very promising thermal biosensors produced on silicon wafers (Fig. 20). Here, a small reactor cell (5 × 1 × 0.014 mm) was prepared on a silicon chip (14 × 6 × 0.4 cm) with minimal heat capacity. The cell consists of 33 parallel channels and has a total volume of 0.02 µl. Microbead thermistors were fixed on gold tubes at the inlet and at the outlet of the silicon chip. The immobilization of enzymes are performed via glutaraldehyde activation of the silicon chip. Enzyme solutions are layered over the channel section and dried at room temperature. The whole preparation has a glass cover attached by a thin layer of silicon rubber glue. Due to the very small reactor cell volume, the residence time of samples in the microchannels is short. Thus, heat leakage from the cell is reduced. Nevertheless, the sensor has to be placed in an aluminium box filled with polyurethane foam. Moreover, huge Wheatstone-bridge equipment is necessary. In comparison to the commonly used ET, the

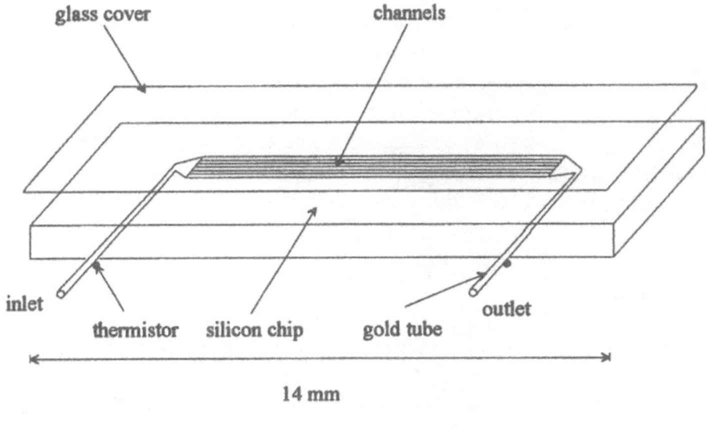

Fig. 20. Miniaturized ET-chip

miniaturized thermal biosensor needs much lower sample volumes, about 20 µl, making it very attractive for clinical use.

5.3 Integrated Thermopiles

Thermocouples are alternative transducers for converting the heat produced by a corresponding enzymatic reaction into an electrical signal. Here, the temperature difference between the ends of a pair of dissimilar metal wires is deduced from a measurement of the difference in the thermoelectric potentials developed along the wires (Seebeck effect). A temperature gradient in a metal or alloy leads to an electrical potential gradient being set up along the temperature gradient. For small temperature differences, the thermoelectric potential gradient ΔV is proportional to the temperature gradient ΔT and strongly dependent on the materials which are used to form the junction. The proportional factor is the relative Seebeck coefficient α_{AB} between materials A and B. Thermocouples have a high rejection ratio for common-mode thermal noise. Moreover, they do not require a well-matched reference, as thermistor based systems. As a consequence, they are not large and suitable for miniaturization.

Bataillard et al. (1993) described the integration of a series of thermocouples (thermopiles) on a silicon chip in order to increase the voltage output. The chip consists of an array of p-type silicon/aluminium thermocouples, connected in series and integrated in a n-type silicon epoxy layer grown on the silicon wafer. The overall size of the chip is only 5 × 5 mm. Several enzymes are immobilized on the array, and the chips are inserted into a FIA system. Glucose is monitored with coimmobilized glucose oxidase and catalase ranging from 2 to 100 mmol/l, urea (1–1000 mmol/l) with immobilized urease and penicillin (1–1000 mmol/l) with immobilized β-lactamase.

Xie et al. (1994) manufactured a microbiosensor with an integrated thermopile on a quartz chip because of the lower heat conductivity of quartz compared to silicon. The size of the whole sensor was 25.2 × 14.8 × 0.6 mm, and immobilized enzymes (CPG beads) were were placed in the chips microchannels. The sensor was applied to glucose analysis in the 2 to 25 mmol/l range, using only 1 µl samples. Due to clinical analysis requirements, the sensor might be interesting for blood glucose measurements.

Nevertheless, some problems have to be considered. First of all, it is very difficult to obtain non-clogging immobilization materials. Moreover, the sensitivity is strongly limited by electrical noise due to the high impedance of the integrated thermopile. Thus, further research and development is necessary, especially as far as the electronics are concerned.

5.4 Bio-Thermochips

Recently, a new ET system was presented which does not require a precise thermostat (Shimohigoshi et al. 1995). The "bio-thermochip" consists of a bead thermistor directly surrounded by an enzyme immobilization support to reduce the heat loss to the surroundings, to enhance the enzyme loading, and to physically protect the enzyme layer (Fig. 21). Nonspecific temperature changes are obtained via a reference thermistor. Complicated thermostating procedures are not necessary. Bio-thermochips with coimmobilized glucose oxidase and catalase were placed in a polystyrene insulated box, and after 5 minutes of temperature equilibration, glucose solutions of 1 to 4 mmol/l were injected through a sample inlet.

The most interesting point of the study is the possibility of not needing thermo-stats due to direct contact between the enzyme and the thermistor bead. However, the insulation box is still very large – 300 × 260 × 175 mm. Moreover, direct enzyme immobilization on thermistors is very expensive because the whole sensing part has to be exchanged after the utilization of the enzymes.

5.5 Compact Multichannel Enzyme Thermistors

Miniaturization of biosensors is of great importance in the field of clinical analysis. However, in the case of bioprocess monitoring and control, small set-ups are not necessary. What bioprocess engineers need are robust devices within stable boxes, simple operating methods and data transfer protection. Moreover, there must be the possibility of exchanging exhausted enzymes very quickly. Within this scope, we continued the development on the whole successful multi-channel ET. The system consists of a 19″-box for industrial applications containing a thermostated aluminium cylinder, an A/D-board for data acquisition and a microcontroller for coordination of the electronic modules. Thermistors resistances are measured via a multiplexer and constant current supply. Thus, the drifting wheatstone bridges are not needed. The compact ET is easily programmed via three buttons and a LED display. The digital data are transferred via RS232 to a PC where complex analysis is performed. Due to digi-

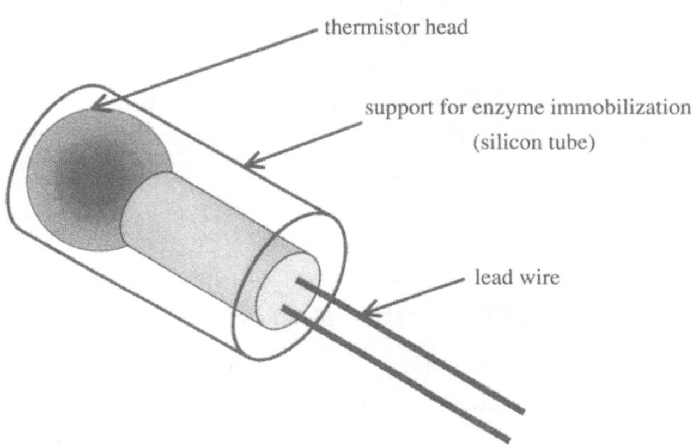

Fig. 21. Sensing part of the bio-thermochip

tal data transfer, the analysis is protected against interference such as that caused by voltage spikes.

Although exhausted enzymes can be replaced quickly, the compact device does not solve the problems of enzyme instability or complex thermostating. Its construction is more practically orientated. The major aim of the study was the transformation of an ET into a more practical useful arrangement.

6 Conclusions

In summary, thermal biosensors are still attracting great interest in several areas of biotechnology due to their wide applications. They will surely not solve all bioanalytical problems. In particular, the TELISA systems are in strong competition with the better working fluorescence assays. Of course, this is not just a reason of the transducer technique but depends on the necessary and expensive conjugate preparations. Nevertheless, in special areas thermal biosensors still have superior characteristics in comparison to other transducer types. When miniaturized, they might be very interesting for clinical glucose monitoring due to their functioning without interference from ascorbic acid. The latter is a big problem for amperometric transducers. Thermal biosensors give their best results with deaminating (e.g., urease, glutaminase, asparaginase) and hydrolytic enzymes (e.g., α-chymotrypsin, pig liver esterase). Since the 1990s, new applications for the ET have been investigated and have enlarged the possibilities of biosensor application. If one summarizes the experiences of the last twenty years, the ET seems to be a very interesting tool for learning basic principles and setting trends in biosensorics.

As mentioned above, miniaturization is not required for bioprocess monitoring. However, the experiences with microcontrollers and constructions of practically useful arrangements might be very efficient in combination with the miniaturized ET and thermopiles developed so far. The construction of devices which are simple to operate should be the target and trend for the next years. The technical prerequisites are: compact boxes containing the whole measuring electronics and the sensor chips with immobilized enzymes that are easy to exchange after exhaustion. In this case, thermal biosensors will have a good chance in biotechnology. However, each analysis problem probably has its own characteristics and several (bio)sensor types should carefully considered.

7 References

Akertek E, Tarhan L (1995) Characterization of immobilized catalases and their application in pasteurization of milk with H_2O_2. Appl Biochem Biotech 50: 291–303

Alonso A, Almendral MJ, Baez MD, Porras MJ, Alonso C (1995) Enzyme immobilization on an epoxy matrix. Determination of L-Arginine by flow-injection analysis. Anal Chim Acta 308: 164–169

Amici AM, Minghetti A, Scotti T, Spalla C, Tognolli L (1967) Ergotamine production in submerged culture and physiology of Claviceps purpurea. Apll Microbiol 15: 597–602

Bataillard P, Steffgen E, Haemmerli S, Manz A, Widmer HM (1993) An integrated silicon thermopile as biosensor for the thermal monitoring of glucose, urea and penicillin. Biosens Bioelect 8:89–98

Bergmeyer UH (1983) Methods of enzymatic analysis. 3. Auflage, VCH Verlagsgesellschaft mbH, Weinheim

Birnbaum S, Bülow L, Hardy K, Danielsson B, Mosbach K (1986) Automated thermometric enzyme linked immunoassay of human proinsulin produced by *Escherichia coli*. Anal Biochem 158:12–19

Brandes W, Maschke HE, Scheper T (1993) Specific flow injection sandwich binding assay for IgG using protein A and a fusion protein. Anal Chem 65:3368–3371

Cannings LM, Carr PW (1975) Rapid thermochemical analysis via immobilized enzyme reactors. Anal Lett 8(5):359–367

Danielsson (1995) personal communication. University of Lund, Sweden

Danielsson B, Bülow L, Lowe CR, Satoh I, Mosbach K (1981b) Evalutaion of the enzyme thermistor as a specific detector for chromatographic procedures. Anal Biochem 117:84–93

Danielsson B, Mosbach K (1988) Enzyme thermistors. Meth Enzymol 137:181–197

Danielsson B, Mattiasson B, Mosbach K (1981a) Enzyme thermistor applications and their analytical applications. Appl Biochem Bioeng 3: pp 97–143

Danielsson B, Mosbach K (1979) Determination of enzyme activities with the enzyme thermistor unit. FEBS Lett 101(1), pp 47–50

Danielsson B, Larsson PO (1990) Specific monitoring of chromatographic procedures. Trends Anal Chem 9:223

Danielsson B, Bülow L, Lowe CR, Satoh I, Mosbach K (1981b) Evalutaion of the enzyme thermistor as a specific detector for chromatograhic procedures. Anal Biochem 117: 84–93

Danielsson B (1994) Enzyme thermistors for food analysis. In: Wagner G, Guilbault GG (eds), Food biosensor analysis. Marcel Dekker, New York, Basel, Hong Kong, pp 173–190

Docolomansky P, Gemeiner P, Mislovicova D, Stefuca V, Danielsson B (1994) Screening of concanavalin A-bead cellulose conjugates using an enzyme thermistor with immobilized invertase as the reporter catalyst. Biotechn Bioeng 43:286–292

Fischer W (1989) Diploma-thesis. University of Hannover, Germany

Flygar L, Larsson P.-O, Danielsson B (1990) Control of an affinity purification procedure using a thermal biosensor. Biotechnol Bioeng 36:723–726

Gemeiner P, Stefuca V, Welwardova, A, Michalkova E, Welward L, Kurillova L, Danielsson B (1993) Direct determination of the cephalosporin transforming activity of immobilized cells with use of an enzyme thermistor. 1. Verification of the mathematical model. Enzyme Microb Technol 15:50–56

Gemeiner P, Stefuca V, Welwardova-Vikartovska A (1996) Screening and design of immobilized biocatalysts through the kinetic characterization by flow microcalorimetry. In: Wijfels RH, Buitelaar RM, Bucke C, Tramper J (eds) Immobilized Cells: Basics and Applications. Elsevier Science BV, Amsterdam, pp 320–327

Grau C (1993), PhD-thesis, University of Hannover, Germany

Guilbault GG, Danielsson B, Mandenius CF, Mosbach K (1983) A comparison of enzyme electrode and thermistor probes for assay of alcohols using alcohol oxidase. Anal Chem 55: pp 1582–1585

Hundeck (1988) diploma-thesis, University of Hannover, Germany

Hundeck HG, Sauerbrei A, Hübner U, Scheper T, Schügerl K (1990) Four-channel enzyme thermistor system for process monitoring and control in biotechnology. Anal Chim Acta 238:211–222

Hundeck HG, Weiß M, Scheper T, Schubert F (1993) Calorimetric biosensor for the detection and determination of enantiomeric excesses in aqueous and organic phases. Biosens Bioelectron 8:205–208

Hundeck HG (1991) PhD-thesis, University of Hannover, Germany

Hundeck HG, Hübner U, Lübbert A, Scheper T, Schmidt J, Weiß M, Schubert F (1992) Development and application of a four-channel enzyme thermistor system for bioprocess control.

In: Scheller F, Schmid RD (eds), Biosensors: Fundamentals, Technologies and Applications. GBF-Monographs 17: 322–330
Jeong YH, Wang SS (1995) Role of glutamine in hybridoma cell culture: effects on cell growth, antibody production, and cell metabolism. Enz Microb Techn 17:47–55
Kirstein D, Danielsson B, Scheller F, Mosbach K (1989) Highly sensitive enzyme thermistor determination of ADP and ATP by multiple recycling enzyme systems. Biosensors 4:231–239
Lammers F, Scheper T (1996) On-line monitoring of enzyme catalyzed biotransformations with biosensors. Enz Microb Techn, submitted
Lammers F (1996) PhD-thesis, University of Hannover, Germany
Mattiasson B, Rieke E, Munnecke D, Mosbach K (1979) Enzyme analysis of organophosphate insecticides using an enzyme thermistor. J Solid-Phase Biochem 4:263–270
Mattiasson B, Danielsson B, Hermansson C, Mosbach K (1977b) Enzyme thermistor analysis of heavy metal ions with use of immobilized urease. FEBS Lett 85:203–206
Mattiasson B, Danielsson B (1982) Calorimetric analysis of sugars and sugar derivatives with aid of an enzyme thermistor. Carbohydr Res 102:273–283
Mattiasson B, Danielsson B, Winquist F, Nilsson H, Mosbach K (1981) Enzyme thermistor analysis of penicillin in standard solutions and fermentation broth. Appl Environm Microbiol 41(4):pp 903–908
Mattiasson B, Mandenius CF, Axelson JP, Danielsson B, Hagander P (1983) Computer control of fermentations with biosensors. Ann NY Acad Sci 413:193–196
Mattiasson B, Borrebaeck C, Sanfridson B, Mosbach K (1977) Thermometric enzyme linked immunosorbent assay: TELISA. Biochim Biophys Acta 483:pp 221–227
Mattiasson B, Mosbach K, Svensson A (1977a) Application of cyanide metabolizing enzymes to environmental control. Enzyme thermistor assay of cyanide using immobilized rhodanese and injectase. Biotech Bioeng 19:1643–1651
Mecklenburg M, Lindbladh C, Hongshan L, Mosbach K, Danielsson B (1993) Enzymatic amplification of a flow-injected thermometric enzyme-linked immunoassay for human insulin. Anal Biochem 212:388–393
Mosbach K, Danielsson B (1974) An enzyme thermistor. Biochim Biophys Acta 364:140–145
Mosbach K, Danielsson B (1981) Thermal bioanalyzer in flow streams – enzyme thermistor devices. Anal Chem 53(1): 83A–84A, 86A, 89A–91A, 94A
Mosbach K (1991) Thermal biosensors. Biosens Bioelectr 6:179–182
Pschyrembel W (Hrsg) (1993) Klinisches Wörterbuch. de Gruyter Berlin, New York
Rank M, Gram J, Stern-Nielsen K, Danielsson B (1995) On-line monitoring of ethanol, acetaldehyde and glycerol during industrial fermentations with Saccharomyces cerevisiae. Appl Microbiol Biotechnol 42:813–817
Rank M, Gram J, Danielsson B (1993) Industrial on-line monitoring of penicillin V, glucose and ethanol using a split-flow modified thermal biosensor. Anal Chim Acta 281:521–526
Saleemuddin M, Husain Q (1991) Concanavalin A: A useful ligand for glycoenzyme immobilization – a review. Enz Microb Technol 13:290–295
Satoh I, Danielsson B, Mosbach K (1981) Triglyceride determination with use of an enzyme thermistor. Anal Chim Acta 131: pp 255–262
Satoh I, Ishii T (1988) Flow-injection determination of inorganic pyrophosphate with use of an enzyme thermistor containing immobilized inorganic pyrophosphatase. Anal Chim Acta 214:409–413
Satoh I (1991a) An apoenzyme thermistor microanlysis for zinc(II) ions with use of an immobilized alkaline phosphatase reactor in a flow system. Biosens Bioelectr 6:375–379
Sauerbrei A (1988) PhD-thesis, University of Hannover, Germany
Scheller F, Schubert F, (1989) Biosensoren. Birkhäuser Verlag Basel, Boston, Berlin
Scheller F, Siegbahn N, Danielsson B, Mosbach K (1985) High sensitivity thermistor determination of l-lactate by substrate recycling. Anal Chem 57:1740–1743
Schmidt HL, Krisam G, Grenner G (1976) Microcalorimetric methods for substrate determinations in flow streams with immobilized enzymes. Biochim Biophys Acta 429:283–290
Shimohigoshi M, Yokoyama K, Karube I (1995) Development of a bio-thermochip and its application for the detection of glucose in urine. Anal Chim Acta 303:295–299

Stefuca V, Gemeiner P, Kurillova L, Danielsson B, Bales V (1990) Application of the enzyme thermistor to the direct estimation of intrinsic kinetics using the saccharose-immobilized invertase system. Enzyme Microb Technol 12:830–835

Stefuca V, Welwardova A, Gemeiner P, Jakubova A (1994) Application of enzyme flow microcalorimetry to the study of microkinetic properties of immobilized biocatalyst. Biotechnol Tech 8:497–502

Svensson A, Hynning PA, Mattiasson B (1979) Application of enzymatic processes for monitoring effluents. Measurements of primary amines using immobilized monoamine oxidase and the enzyme thermistor. J Appl Biochem 1:318–324

Urban G, Kamper H, Jachimowicz A, Kohl F, Kuttner H, Olcaytuc F, Pittner F, Schalkhammer T, Mann-Buxbaum E (1991) The construction of microcalorimetric biosensors by use of high resolution thin-film thermistors. Biosens Bioelect 6:275–280

Winquist F, Danielsson B, Malpote J.-Y, Persson L, Larsson M-B (1985) *E*nzyme thermistor determination of oxalate with with immobilized oxalate oxidase. Anal Lett 18:573–588

Xie B, Danielsson B, Norberg P, Winquist F, Lundström I (1992) Development of a thermal micro-biosensor fabricated on a silicon chip. Sens Act B 6:127–130

Xie B, Hedberg U, Mecklenburg M, Danielsson B (1993) Fast determination of whole blood glucose with a calorimetric micro-biosensor. Sens Act B 15–16, 141–144

Xie B, Mecklenburg M, Danielsson B, Öhmann O, Winquist F (1994) Microbiosensor based on an integrated thermopile. Anal Chim Acta 299:165–170

Investigation of Catalytic Properties of Immobilized Enzymes and Cells by Flow Microcalorimetry

Vladimír Štefuca [1] · Peter Gemeiner [2]

[1] Department of Chemical and Biochemical Engineering, Slovak University of Technology, Radlinského 9, SK-812 37 Bratislava, Slovak Republic. *E-mail: stefuca@cvt.stuba.sk*
[2] Institute of Chemistry, Slovak Academy of Sciences, Dúbravská cesta 9, SK-842 38 Bratislava, Slovak Republic. *E-mail: chempege@savba.sk*

The investigation of catalytic properties of immobilized biocatalysts (IMB) is a time-consuming and not-always-simple procedure, requiring a simple and accurate method of enzyme-activity measurement. In comparison with generally-used techniques, flow microcalorimetry (FMC) has proven to be a very practical and versatile technique for direct monitoring of the course of enzyme reactions. The principal advantage of FMC is integration of the enzyme reaction and its monitoring in one step. This review summarizes the information needed for the complete kinetic or catalytic characterization of the IMB by FMC, without the requirement of any independent analytical method. The optimal experimental procedure is proposed. Examples of experimental studies on immobilized biocatalysts using the FMC are provided. The method is applicable to purified enzymes as well as to enzymes fixed in cells.

Keywords: Flow microcalorimetry, Immobilized biocatalysts, Reaction rate monitoring, Investigation of kinetic properties, Characterization of bioaffinity systems.

Advances in Biochemical Engineering / Biotechnology, Vol. 64
Managing Editor: Th. Scheper

List of Symbols and Abbreviations

A	Total area of immobilized enzyme particles
c_S, c_P	Particle substrate and product concentration, respectively
c_{Sb}, c_{Pb}	Bulk phase substrate and product concentration, respectively
C_P	Specific heat capacity of liquid phase
D_S, D_P	Effective diffusion coefficient of substrate and product, respectively
ΔH_r	Molar reaction enthalpy
k_1	First order rate constant
K_m	Michaelis constant
K_i	Inhibition constant
P	Parameter in Eq. (29)
$\mathbf{P}$	Vector of kinetic parameters
r	Particle radial coordinate
R	Particle radius
T	Absolute temperature
ΔT_r	Temperature difference between column input and output
v_0	Initial reaction time
v_{kin}	Kinetic reaction rate corresponding to insignificant mass transfer limitation
v_{obs}	Observed reaction rate
V_C	Total bed volume in column
V_L	Bed void volume in column
V_m	Maximum reaction rate
V_p	Total particle volume in column
V_T	Total liquid volume in recirculation system
w	Superficial flow rate
z	Axial coordinate
Z	Length of packed bed
α	Transformation parameter defined by Eq. (18)
ε	Bed void fraction
Φ	Thiele modulus
η	Effectiveness factor
ϱ	Fluid density
Con A	Concanavalin A
CPG	Controlled pore glass
ELISA	Enzyme-linked immunoassay
FMC	Flow microcalorimetry
IMB	Immobilized biocatalyst

1 Introduction

The immobilized biocatalyst (IMB) is a key component of biotransformation systems that are used to transform substrates to desired products. The improvement of biocatalyst properties has a direct influence on the overall effectiveness of the process based on the biotransformation. The basic catalytic characteristics of biocatalyst that are followed include kinetic properties, pH optima, stability, and inhibition. The investigation of catalytic properties of immobilized enzymes is still a time consuming procedure and is not always simple. In the 1980s, a major effort was made to standardize the rules by which IMB is characterized. The Working Party of EFB on immobilized biocatalysts has formulated principles of individual methods, among them the requirement of kinetic characterization [1]. It was recommended to use a packed-bed reactor, equipped with temperature control and with infinite flow circulation. The system should be equipped with a post-column unit to measure the time-dependence of the product or substrate concentration [2, 3], the most commonly used analytical methods being spectrophotometry, chemiluminiscence, automatic titration, bioluminiscence, chromatography, polarimetry, and biosensors based on the oxygen electrode. There are two main drawbacks to the application of these methods:

1. The need to vary the analytical principles, depending on the chemical and physical-chemical properties of analytes;
2. In some cases, mainly in the study of hydrolytic enzymes, the natural substrate must be replaced by an artificial one, that is chromolytic, chromogenic, chemiluminiscent, bioluminiscent, or fluorescent.

Therefore, in the same period, there was great interest in the development of a standard measurement technique, that would join the advantages of immobilized enzymes with a universal detection principle, that lead to a variety of devices for flow enthalpimetry [4]. The first such effort considered the measurement of glucose concentration using immobilized glucose oxidase [5]. An important improvement of this measurement system was the integration of thermistors with immobilized enzyme columns [6] and further technical modifications [7–13]. The research on enthalpimetric methods was motivated by the potential use of enthalpimeters in bioanalytical chemistry [14]. It was impelled by the need to replace complicated and expensive commercial equipment by a system, that would be simpler, cheaper, and have a shorter response time [15]. These requirements were best fulfilled by the concept of an "enzyme thermistor", developed by Danielsson [16]. The basic design and further development of this equipment was aimed at providing a system of flow-injection analysis of the concentration of metabolites, inhibitors [17–19], lipids [20], antigens [21], and of on-line monitoring of fermentation processes and biotransformations [22].

In addition to the analytical applications, there was sporadic work on the employment of flow calorimetry for the investigation of enzyme kinetics [23, 24]. In 1985 Owusu et al. [25] published the first report on the use of flow microcalorimetry for the study of immobilized enzyme kinetics approaching fulfilment of the requirements mentioned above. Later, several papers were published that developed the methodology for the characterization of the catalytic properties of immobilized enzymes [26–33]. Flow microcalorimetry proved to be very practical and versatile technique. The principal advantage of FMC is the integration in one step of the enzyme reaction and monitoring of the reaction course. In spite of its evident advantages, enzyme flow microcalorimetry is not very widely used. One of the reasons for this may be the relatively wide dispersion of published data on possible applications. Another reason is the nature of calorimetric measurement: The thermometric measurement does not provide direct information on changes in the concentration of the reacting compounds; therefore the results of the measurement need to be converted in order to estimate the reaction rate. The authors of this review believe that the broad application of FMC in routine analysis of the properties of immobilized biocatalysts can be facilitated by providing systematic information about the methodology required for manipulation of this technique. In this review, simple mathematical models useful for FMC data treatment are presented, and methods for facilitating the interconversion of thermometric and rate data are presented. Several examples of the application of the technique are provided.

2 Equipment and Procedure

Figure 1 depicts the experimental arrangement that is in regular use in our laboratories for the investigation of the properties of IBM by flow microcalorimetry. The main functional part of the equipment is the flow microcalorimeter 3300 Thermal Assay Probe (Advanced Biosensor Technology AB Lund, Sweden). It consists of a thermostated block containing a small column packed with an IMB (FMC unit), probes equipped with thermistor sensors connected to a measurement and control unit, i.e., a Wheatstone bridge and temperature control. The analog signal provided by FMC unit is digitized and registered by a personal computer. The system is equipped with a peristaltic pump and a switching valve that selects between an open mode (all material from the column output is wasted) and a closed mode (total flow circulation is achieved). The principle of measurement is registration of the temperature change induced by the heat of

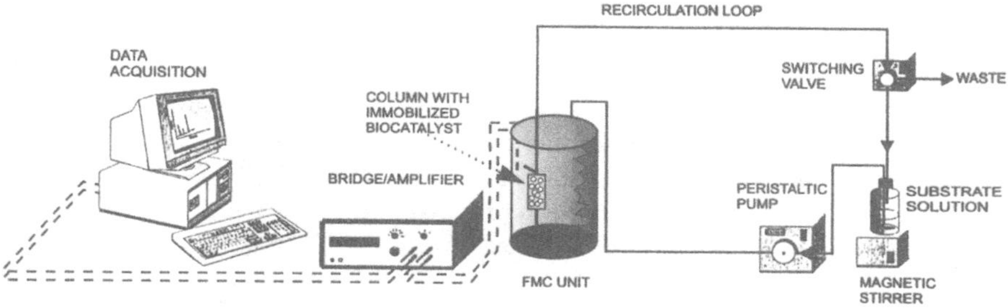

Fig. 1. Experimental setup for the enzyme flow microcalorimetry

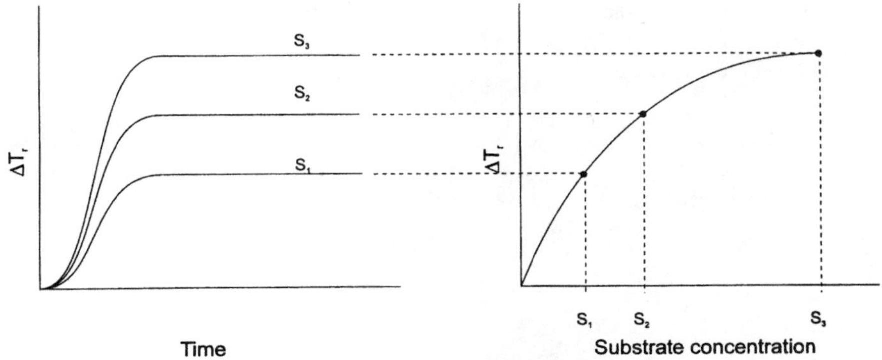

Fig. 2. Steady-state heat response measurement for different input substrate concentrations

the enzyme reaction released in the column (standard dimensions of the packed bed are 2 cm length and 0.4 cm inner diameter). The column is operated as a packed bed reactor. At the beginning of experiment the system is stabilized by pumping a buffer solution through the column. After thermostating, the measurement is begun by replacing the buffer with the substrate solution.

The most common procedure is pumping a substrate solution of a defined composition (concentration of substrate and other compounds having influence on the enzyme activity) through the microcalorimetric column. The basic information provided by the microcalorimetric measurement is the relation between reaction conditions and the steady-state heat response, ΔT_r, measured as the temperature difference between the column input and output. Figure 2 is an illustration of such measurement. In the next part of this review, the mathematical assessment of the experimental data, based on mass and heat balances, is provided.

3 Heat and Material Balance

The value of a thermometric signal is calculated by means of a mathematical formulation of the heat and material balance in the reaction system, represented by the microcalorimetric column packed with an immobilized enzyme. The microcalorimetric column with an IMB can be defined as the continuous packed bed reactor depicted in Fig. 3. The balance equations were derived according to the following simplifying assumptions, which had been previously verified [27]:

- changes of the substrate concentration and temperature along the reactor are so small that the reaction rate, the molar reaction enthalpy and the fluid specific heat capacity are constant,
- plug flow occurs in the reactor;
- the flow rate is sufficiently high to prevent reaction-rate limitation by external mass transfer,
- heat losses from the reactor are neglected and the reactor is considered to be adiabatic.

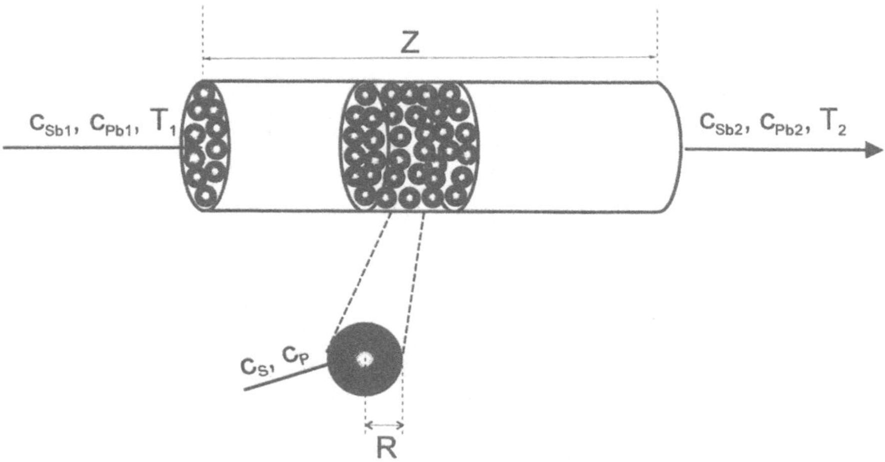

Fig. 3. The reactor model describing the column packed with particles of immobilized biocatalyst

When no heat losses cross the reactor wall are present, the local temperature change depends directly on the reaction rate. Then, material and heat balances can be defined as

$$-w\frac{dc_{Sb}}{dz} = \varepsilon\eta v_{kin} \tag{1}$$

$$w\frac{dc_{Pb}}{dz} = \varepsilon\eta v_{kin} \tag{2}$$

$$w\rho C_P\frac{dT}{dz} = \varepsilon\eta v_{kin}(-\Delta H_r) \tag{3}$$

where c_{Sb} and c_{Pb} are the local bulk-phase concentrations of substrate and product, respectively, T is the absolute temperature, z is the axial coordinate, η is the effectiveness factor expressing the effect of the particle mass transfer on the reaction rate. V_{kin} is the intrinsic reaction rate considered as the reaction rate per unit of void volume of the packed bed. The superficial flow rate, w, bed void fraction, ε, fluid density, ρ, molar reaction enthalpy, ΔH_r, and fluid specific heat capacity, C_p, will be considered constant for the specific experimental system.

The above differential equation system is coupled with one of boundary conditions according to the notation used in Fig. 3:

$$z = 0: \quad T = T_1 \quad c_{Sb} = c_{Sb1} \quad c_{Pb} = c_{Pb1} \tag{4}$$

$$z = Z: \quad T = T_2 \quad c_{Sb} = c_{Sb2} \quad c_{Pb} = c_{Pb2}. \tag{5}$$

3.1 Mass Balance in Immobilized Enzyme Particle

The effectiveness factor in Eqs. (1)–(3) is defined as

$$\eta = \frac{v_{obs}}{v_{kin}} \tag{6}$$

v_{kin} represents the reaction rate in the kinetic region, it means the rate that would be attained if the substrate concentration in the whole particle were the same as that in the bulk phase, c_{Sb}. In order to calculate the observed reaction rate, v_{obs}, the particle mass balance equations have to be solved:

$$\frac{D_S}{r^n}\frac{d}{dr}\left(r^n \frac{dc_S}{dr}\right) = v(c_S, c_P, \mathbf{P}) \tag{7}$$

$$\frac{D_P}{r^n}\frac{d}{dr}\left(r^n \frac{dc_P}{dr}\right) = -v(c_S, c_P, \mathbf{P}) \tag{8}$$

while required boundary conditions are

$$r = 0: \quad \frac{dc_S}{dr} = \frac{dc_P}{dr} = 0 \tag{9}$$

$$r = R: \quad c_S = c_{Sb} \quad c_P = c_{Pb}. \tag{10}$$

The kinetic term, v, is a function of the kinetic parameters: vector **P** and the particle substrate and product concentrations, c_S and c_P, respectively. D_S and D_P are the corresponding effective diffusion coefficients and r is the particle coordinate (in the case of spherical geometry it is the radial distance). Parameter n depends on the geometry of the biocatalyst particle and is 0, 1, 2 for a plate, a cylinder and a sphere, respectively. Since concentrations on the particle surface are assumed to be identical with bulk concentrations, boundary conditions do not include the influence of external mass transfer. Solving the above differential equations, the observed reaction rate in the packed bed is evaluated from the rate of substrate flux to the particle or of product flux from the particle

$$v_{obs} = \frac{A \cdot D_S}{V_L}\left(\frac{dc_S}{dr}\right)_{r=R} = -\frac{A \cdot D_P}{V_L}\left(\frac{dc_P}{dr}\right)_{r=R} \tag{11}$$

where A is the total external surface area of the particle and V_L is the volume of the bulk phase in the packed bed. For spherical geometry, Eq. (11) can be simplified to

$$v_{obs} = \frac{(1-\varepsilon)}{\varepsilon R}\left(\frac{dc_S}{dr}\right)_{r=R} = -\frac{(1-\varepsilon)}{\varepsilon R}\left(\frac{dc_P}{dr}\right)_{r=R} \tag{12}$$

The kinetic rate, v_{kin}, is expressed by the relation

$$v_{kin} = \frac{V_P}{V_l} v(c_S, c_P, P) \tag{13}$$

or taking into acount the void fraction of packed bed this is

$$v_{kin} = \frac{1-\varepsilon}{\varepsilon} v(c_S, c_P, P). \tag{14}$$

It follows from the form of the model equations used, the temperature profile in the particle is not considered in the calculation of the observed reaction rate, because – under steady-state conditions – no heat accumulation occurs in the biocatalyst particle. Consequently, the variation of reaction rate with temperature change can be neglected, in view of the low temperature differences typical for enzyme flow microcalorimetry.

3.2
Low Substrate Conversion in the Column

The mathematical description of microcalorimetric data can be simplified when the amount of immobilized enzyme preparation in the column is minimized, so that the change in the reaction conditions due to the progress of the reaction does not influence the reaction rate. This condition is fulfilled when a sufficiently low substrate conversion is achieved in the column. In general, the substrate conversion is determined by analyzing the substrate concentration at the column output by standard analytical techniques. There is, however, a simpler procedure for avoiding the post-column analysis, a procedure based exclusively on measurement with the flow microcalorimeter. Steady-state reaction conditions are established in the flow microcalorimeter in the first experimental run. Then, a sufficient volume of the reaction mixture in the column output is collected and reused as the input flow for the second experimental run. If the difference between the thermometric signals of the first and second experiment is negligible (less than 5%), the reactionrate change along the column can be neglected.

The condition of a small controlled amount of the IMB in the column allows description of the column as a reactor with a differential packed bed, which means that the reaction rate does not change along the bed. Then, the balance Eqs. (1)–(3) take the form of following difference equations:

$$\frac{dc_{Sb}}{dz} = \frac{\Delta c_{Sb}}{\Delta z} = \frac{c_{Sb2} - c_{Sb1}}{Z} = -\frac{\varepsilon}{w} \eta v_{kin} \tag{15}$$

$$\frac{dc_{Pb}}{dz} = \frac{\Delta c_{Pb}}{\Delta z} = \frac{c_{Pb2} - c_{Pb1}}{Z} = \frac{\varepsilon}{w} \eta v_{kin} \tag{16}$$

$$\frac{dT}{dz} = \frac{\Delta T}{\Delta z} = \frac{\Delta T_r}{Z} = \frac{\varepsilon(-\Delta H_r)}{w\rho C_P} \eta v_{kin} \tag{17}$$

where Z is the length of the packed bed. All of the constant parameters can be included in parameter α

$$\alpha = \frac{Z\varepsilon(-\Delta H_r)}{w\rho C_P} \tag{18}$$

and Eq. (17) can be rewritten

$$\Delta T_r = \eta \alpha v_{kin} \tag{19}$$

or

$$\Delta T_r = \alpha v_{obs}. \tag{20}$$

4 Transformation of Thermometric Data to Reaction Rates

The typical feature of the FMC is that the thermometric signal, ΔT_r, and not the true value of reaction rate is measured. In the approach of differential beds, there is, however, a linear relation between ΔT_r and the overall reaction rate, v_{obs} [Eq. (20)]. This means that – if the value of α is known – reaction rate values can be calculated from the measured temperature changes using Eq. (20) in the form

$$v_{obs} = \frac{\Delta T_r}{\alpha} \tag{21}$$

When the experimental data are available as a relation between overall reaction rates and reaction conditions, they can be treated by procedures based on the solution of mass balance equations, and the kinetic parameters can be determined, for example, by regression methods. The basic task that remains is to determine the value of the parameter α. There are three possibilities to do this:

- calculation from its definition [Eq. (18)],
- calibration based on the investigation of the relation between ΔT_r and the reaction rate determined by the post-column analysis of the concentration of one of the reactants or by the measurement of the reaction rate in an independent reaction module [28],
- the microcalorimeter autocalibration procedure explained below.

The disadvantage of the first approach is the requirement to know the quantities involved in the energy balance (e.g. molar reaction enthalpy, fluid heat capacity etc.). This disadvantage is eliminated by approaches based on calibration of the microcalorimeter. Moreover, calibration can compensate for small systematic errors produced by the microcalorimetric equipment, as well. In the second approach, however, the post-column analysis of reactant concentrations by an independent analytical technique still complicates the measurement and makes it more time consuming. In addition, in many reactions substrate conversion in the FMC column is too low to be measured with sufficient precision by conventional analytical techniques. Any enhancement of the substrate conversion by decreasing the volumetric flow rate increases external mass transfer limitations and heat losses. This complicates the mathematical description necessary for the estimation of kinetic parameters. Therefore, it was proposed that the activity of the immobilized biocatalyst can be determined more conveniently by carrying out an additional reaction rate measurement outside the FMC column [28].

The disadvantages of the previous techniques provided the motivation to develop another approach: an autocalibration technique. This method does not require either an independent analytical method for calibration, post-column analysis, or separate rate determination.

The novel feature of the method is the modification of experimental setup to apply total recirculation of the reaction mixture, as depicted in Fig. 1. By switching the valve from waste to the circulation loop, the system is closed and the steady state – cannot be reached, because of a continuous consumption of substrate in the enzyme reaction. In order to simplify the mathematical description, the following experimental assumptions were made:

1. perfect mixing in the stirred reservoir,
2. sufficiently high volumetric flow rate in the system in relation to the reaction rate in the column so that the concentration difference between the reservoir and the column can be neglected.

Conditions (1) and (2) imply that, from the point of view of the mathematical description, the circulating system will behave as a stirred batch reactor. Then, the initial rate of reaction, v_0, in the reactor is

$$v_0 = \left(\frac{dc_S}{dt}\right)_{t=0} = \frac{V_C}{V_T \varepsilon} v_{obs} \tag{22}$$

where V_C and V_T are the volume of the packed bed in the column and the total volume of the circulation system, respectively. According to condition (2), the substrate concentration measured in the stirred reservoir, c_S, corresponds to the concentration in the column. At the beginning of the experiment, FMC is operated in the open flow mode. After achieving the steady state (the signal remains stable), at time = 0 min, the column output is shunted into the stirred reservoir and the thermometric signal is registered continuously. An example of the time-variation of the measured signal is depicted in Fig. 4B.

Considering the differential packed bed and the slow dynamics of the reaction system, a pseudo-steady-state in the column can be assumed for a short period of time, in spite of the non-steady-state in the whole system. This assumption is valid when the total volume of liquid in the circulation system is sufficiently large compared to the volume of the column and the flow rate is so high that in the time interval equal to the column residence time the signal decrease is less than 1%. This means that each thermometric signal corresponds to one value of the substrate concentration, the concentration that would

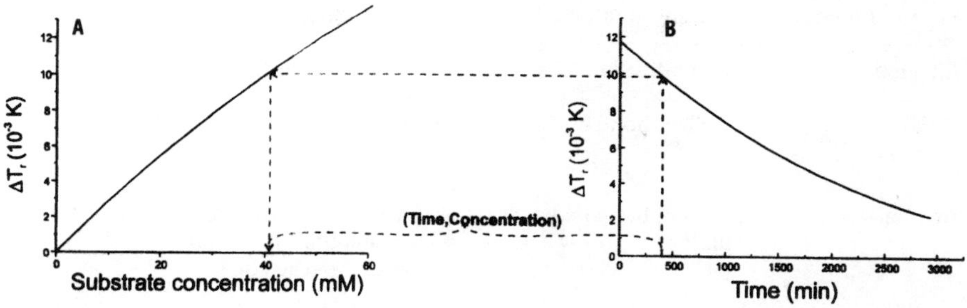

Fig. 4. Conversion of thermometric data

produce the same thermometric signal at steady state. This assumption enables transformation of the signal-time dependence to substrate-concentration-time dependence. Determination of this dependence at one point is indicated in Fig. 4. The calibration dependence shown in Fig. 4A is, in effect, the same as the dependence of the steady state signal on substrate concentration shown in Fig. 2. After establishing the dependence of substrate concentration on time, the reaction rate in the column can be calculated using Eq. (22) in the form

$$v_{obs} = v_0 \frac{V_T \varepsilon}{V_C} \tag{23}$$

where the substrate concentration at time zero, v_0, is derived from the obtained subtrate-time dependence. The determined reaction rate in the column, v_{obs}, is linked to the initial thermometric signal, ΔT_r. Using these values the transformation parameter α is calculated from Eq. (21)

$$\alpha = \frac{\Delta T_r}{v_{obs}} \tag{24}$$

In the following section examples are given that illustrate the application of the principles introduced above to the study of the properties of immobilized biocatalysts.

5
Applications of the Method

5.1
Investigation of Enzyme Kinetics

In spite of the evident advantages of flow microcalorimetry, this technique has been used in only a few experimental investigations of enzyme kinetics. Several references dealing with this subject are listed in Table 1.

Different aspects of the characterization of the properties of IMB were investigated in the literature cited. Section 5 systemizes the available examples in the frame of the methodology proposed by this review.

Table 1. Immobilized biocatalysts studied by the flow microcalorimetry

Enzyme	Immobilization technique	Mathematical model	Ref.
Urease	Covalent bond on CPG	Particle mass transfer, product inhibition kinetics, integral packed bed	[25]
Invertase	Covalent bond on bead cellulose	No particle mass transfer, substrate inhibition kinetics, differential packed bed	[27]
Co-immobilized Glucose oxidase-Catalase	Covalent bond to CPG	Particle mass transfer, product inhibition kinetics, integral packed bed	[26]
Invertase	Covalent bond to porous silica	Qualitative observations without mathematical description	[33]
D-Amino acid oxidase	a) Cells entrapped in Ca-pectate gel	Particle mass transfer, first-order kinetics, differential packed bed	[28]
	b) Cells entrapped in polyacrylamide gel		[28]
Invertase	Biospecific adsorption adsorption on Con A-bead cellulose	No particle mass transfer, substrate inhibition kinetics, differential packed bed	[30, 31]
Penicillin acylase	Cells entrapped in calcium pectate gel	Particle mass transfer, substrate inhibition kinetics, differential packed bed	[29]

5.1.1
Conversion of Thermometric Data

The mathematical treatment of FMC data can be accomplished by standard procedures via the solution of mass balance equations, on condition that the data were converted to reaction rate data with Eq. (21). As mentioned above, this requires the determination of the transformation parameter α. Two approaches based on calibration were developed and tested. In the first approach, thermometric signals are combined with the absolute activity of IMB, which had been determined by a separate measurement using an independent analytical technique. Figure 5 shows a calibration for the cephalosporin C transformation catalyzed by D-amino acid oxidase. The activity of the IMB was determined by the reaction rate measurement in a stirred-tank batch reactor. The reaction rate was determined as the initial rate of consumption of cephalosporin C monitored by HPLC analysis. The thermometric response was measured for each IMB packed in the FMC column, and plotted against the corresponding reaction rate. From the calibration results shown in Fig. 5 it can be concluded, independently of the type of immobilized biocatalyst, that the data fall to the same line and that there is a linear correlation between the heat response and the activity of the catalyst packed in the column. The transformation parameter α was determined from

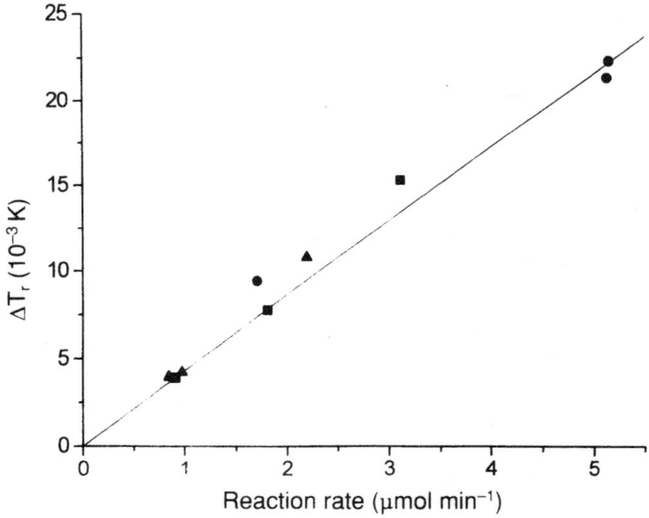

Fig. 5. Correlation between heat response and reaction rate of cephalosporin C transformation by immobilized D-amino acid oxidase of *Trigonopsis variabilis*. Enzyme immobilization techniques: entrapment in polyacrylamide gel (■), cells cross-linked with glutaraldehyde (●), cells entrapped in polyacrylamide gel (▲) [28]

the slope calculated by linear regression and its value was $\alpha = 4.44$ K (μmol min^{-1})$^{-1}$ [28].

Since each calibration point is performed in a newly packed column, there is important data dispersion around the calibration line. In addition, the entire calibration procedure is rather time consuming. Therefore, another approach was developed, that provides more accurate results in a considerably shorter time. This recently published approach [32] employs a slightly modified version of the same FMC equipment. Instead of a flow-through arrangement enabling the measurement in steady state, the measurement is performed in a system with continuous circulation, as shown in Fig. 1.

The method was confirmed experimentally with sucrose hydrolysis catalyzed by invertase, that had been immobilized by biospecific binding on concanavalin A-bead cellulose. Figure 6B shows good agreement between the substrate concentration determined by the conversion of thermometric signals of Fig. 6A and those obtained by spectrophotometric analysis. From the data shown in Fig. 6B, the initial rate was determined to be $v_0 = 0.775$ mM min^{-1}. Introducing this value into Eq. (23), v_{obs} was calculated and the value of the transformation parameter α was determined from Eq. (24).

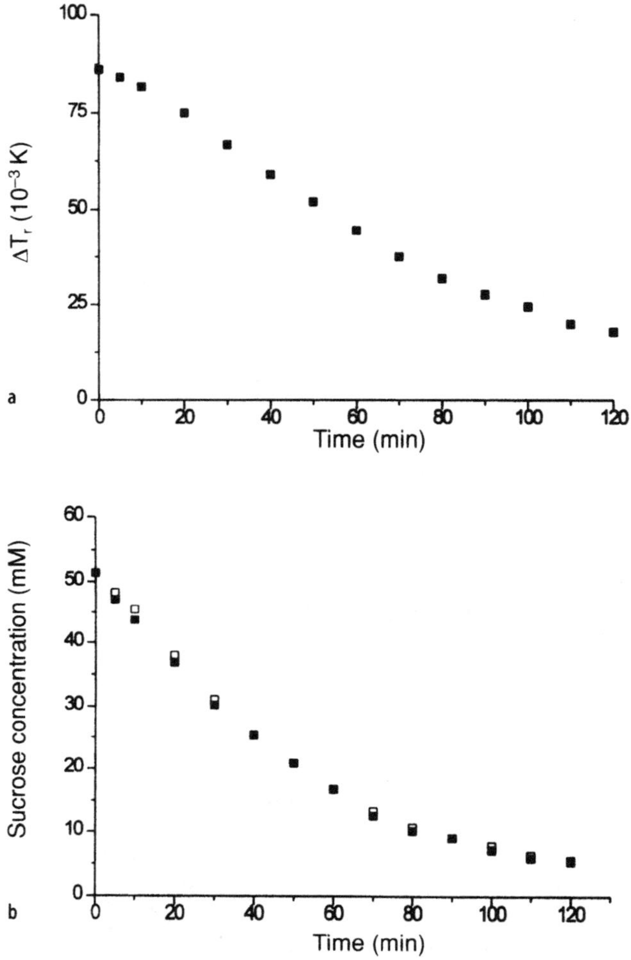

Fig. 6. Investigation of kinetic properties of immobilized invertase by flow microcalorimetry in the circulation mode. Initial sucrose concentration 51 mM, invertase immobilization by biospecific binding on concanavalin A-bead cellulose was prepared by binding on concanavalin A linked to chlorotriazine-activated cellulose. **a** Raw experimental thermometric data; **b** data after conversion by the procedure indicated in Fig. 4. Concentrations were determined spectrophotometrically (*open symbols*) and by transformation of thermometric data explained in Section 5 (*closed symbols*) [32]

5.1.2
Reaction Systems Without Limitation by Particle Diffusion

In many cases the influence of particle mass transfer on the reaction rate can be neglected (value of Thiele modulus approaches zero or the enzyme is immobilized exclusively on the particle surface). Then, value of the effectiveness factor is unity and the solution of Eqs. (1)–(3) is substantially simpler.

This approach was used for the study of the kinetic properties of invertase immobilized on the cellulose bead surface [27, 30, 33]. In the following example the kinetic model described for invertase was used [34]

$$v(c_{Sb}, c_{Pb}, P) = \frac{V_m c_S}{K_m + c_S + \frac{c_S^2}{K_i}} \frac{W}{W_0} \tag{25}$$

where W/W_0 expresses the effect of the decrease in the activity of water on the reaction rate at high sucrose concentrations [34]. Introducing the kinetic model in Eqs. (1), (3) and (14) we obtain

$$\frac{dc_{Sb}}{dz} = -\frac{(1-\varepsilon)}{w} \frac{V_m c_{Sb}}{K_m + c_{Sb} + \frac{c_{Sb}^2}{K_i}} \frac{W}{W_0} \tag{26}$$

$$\frac{dT}{dz} = -\frac{(1-\varepsilon)(-\Delta H_r)}{w \rho C_p} \frac{V_m c_{Sb}}{K_m + c_{Sb} + \frac{c_{Sb}^2}{K_i}} \frac{W}{W_0} . \tag{27}$$

Depending on the overall activity of the enzyme immobilized in the column two characteristic cases were studied:

1. The activity was so low that the change in substrate concentration along the column had no influence on the enzyme reaction rate and the column was regarded as a differential reactor.
2. The activity was so high that the change in substrate concentration along the column could not be neglected.

5.1.2.1
Low Enzyme Activity

In this case, which had been investigated in previous work [27], the results were identical to those of kinetic measurements using an independent analytical method. Introducing the kinetic model [Eq. (25)] into Eq. (14) and combining with Eq. (19), the following expression is obtained:

$$\Delta T_r = \frac{\alpha(1-\varepsilon)V_m}{\varepsilon} \frac{c_{Sb}}{K_m + c_{Sb} + \frac{c_{Sb}^2}{K_i}} \frac{W}{W_0} . \tag{28}$$

The significance of Eq. (28) is that the thermometric data can be used for evaluation of the kinetic parameters K_m and K_i, and, thus, for rapid determination of the kinetic properties of IMB preparations. This had been performed in the previous work by investigating the kinetic properties of immobilized invertase [27]. Typical kinetic data from this study are presented in Fig. 7. In this

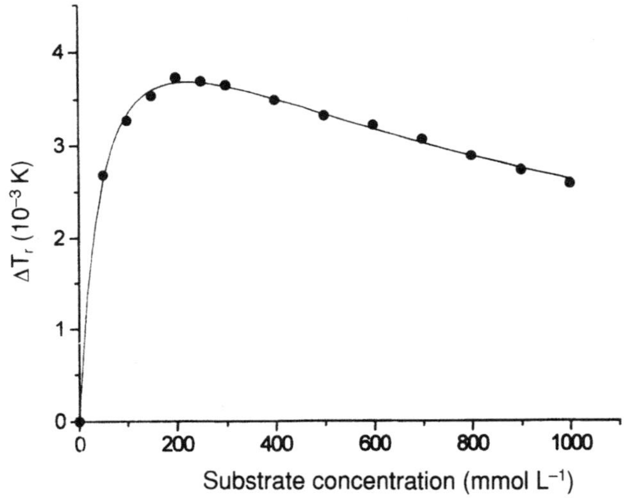

Fig. 7. Effect of sucrose concentration on thermometric signal in steady-state. Invertase was immobilized on bead cellulose activated by cyanuric chloride [27]

situation the maximum reaction rate, V_m, cannot be determined independently from the parameter α. Moreover, the data serve as a relative measure of the comparison of activity among IMB preparations [30].

5.1.2.2 High Enzyme Activity

When a significant change in the substrate concentration in the column is expected, numerical integration of the model equations (26) and (27) should be applied to the experimental data. The solution was simplified by combining Eq. (27) with Eq. (26) in order to obtain the following equation, where the term $\frac{(-\Delta H_r)}{\rho C_p}$ was defined as parameter P and was considered to be constant in the experimental system under investigation:

$$\frac{dT}{dz} = -\frac{dc_{Sb}}{\Delta z}\frac{(-\Delta H_r)}{\rho C_p} = -\frac{dc_{Sb}}{dz}P. \tag{29}$$

The integration of Eq. (29) gives the following algebraic equation:

$$\Delta T_r = (c_{Sb1} - c_{Sb2})P. \tag{30}$$

For the calculation of the thermometric signal value, ΔT_r, the output substrate concentration c_{Sb2} was determined by the integration of Eq. (26). Using this value and the value of the input substrate concentration, c_{Sb1}, the heat response, ΔT_r, was determined from Eq. (30).

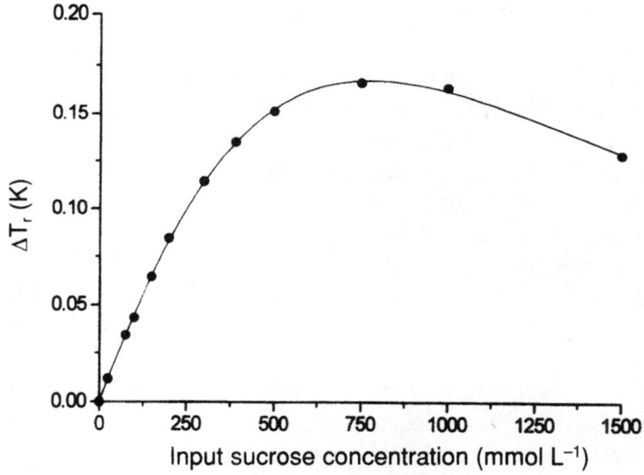

Fig. 8. Flow microcalorimetric investigation of kinetic properties of invertase bound to Eupergit C activated by concanavalin A. The line corresponds to calculated data. The model involving high substrate conversion was used and parameters were estimated by nonlinear regression: $V_m = 1950$ mM min^{-1}, $K_m = 3420$ mM, $K_i = 246$ mM, $P = 2.7$ K (unpubl. results)

The results of the experimental study and mathematical modeling of the invertase-catalyzed hydrolysis of sucrose are displayed in Fig. 8. The analysis of the output substrate concentration showed substantial substrate conversion. Therefore, the data were treated by differential equations (26) and (30), whereas the kinetic parameters were fitted using nonlinear regression. Regardless of the good agreement of the calculated and experimental values, it was concluded on the basis of a comparison of kinetic parameters obtained with those known from previous works on similar preparations of immobilized invertase [30] that this method did not provide reliable results.

In contrast, the measurement of the kinetic properties of immobilized urease under conditions of high substrate conversion resulted in "good" data [25]. Evidently, the increased substrate conversion in the microcalorimetric column increases the change in temperature. Thus, better control of heat loss is necessary. Moreover, the treatment of data of integral reactor is more complicated, and can lead to larger errors of parameter estimates. In spite of these difficulties, it is worth putting more effort into the development of methods based on higher substrate conversion because it could be very useful in the investigation of product inhibition effects and in the case of enzyme reactions with low enthalpy of reaction.

5.1.3
Enzyme Systems Influenced by Particle Diffusion Limitation

Many immobilization techniques provide biocatalysts in which the enzyme is immobilized in the porous structure of the biocatalyst particle. In such cases, the observed biocatalytic reaction rate is influenced by exchange of mass between the interior of the particle and its surroundings. A reaction rate profile is built up inside the particle, and the overall reaction rate is different from the kinetic rate. Therefore, the effectiveness factor is not longer unity and its value must be introduced into the mathematical model. The effectiveness factor is calculated by solving the particle mass balance equations (7) and (8).

The first investigation of the influence of particle mass transfer on the reaction kinetics in a flow microcalorimeter, dealing with properties of urease immobilized on controlled pore glass, was published in 1985 [25]. More recently, the evaluation of microcalorimetric data in the case of particle-diffusion limitation was improved and simplified by introducing the principle of the differential bed [28, 29].

In certain cases, restriction of the experimental conditions to low substrate concentrations ($c_S \ll K_m$) is an acceptable condition for the investigation of biocatalyst properties. In this case, the enzyme kinetics can be simplified to the form of a pseudo-first order kinetics expressed by the relation

$$v(c_S, c_P, \mathbf{P}) = \frac{V_m}{K_m} c_S = k_1 c_S \tag{31}$$

where k_1 is the pseudo-first order rate constant. Then, Eq. (7) can be solved analytically and, for example, in the case of spherical geometry the following explicit expression can be obtained for the effectiveness factor calculation [36]:

$$\eta = \frac{3}{\Phi}\left(\frac{1}{\tanh \Phi} - \frac{1}{\Phi}\right) \tag{32}$$

where Φ is the Thiele modulus

$$\Phi = R\sqrt{\frac{k_1}{D_S}}\,. \tag{33}$$

This approach was used in the study of the properties of D-amino acid oxidase isolated or fixed in cells of *Trigonopsis variabilis* and entrapped in calcium pectate or polyacrylamide gel [28]. The approach of a differential reactor (low enzyme activity in the packed bed) was applied. The experimental thermometric data, ΔT_r, were transformed to reaction rates, v_{obs}, according to Eq. (21), whereas parameter α was determined by the calibration shown in Fig. 5. The data were described by the equation

$$v_{obs} = \eta v_{kin} = \eta \frac{(1-\varepsilon)}{\varepsilon} k_1 c_{Sb} \tag{34}$$

that was obtained by combining Eqs. (6), (14) and (31). The kinetic parameter, k_1, was evaluated by comparing the experimental and calculated data. The results of the treatment of the data, listed in Table 2, illustrate the influence of cell

Table 2. Intrinsic rate constant and effectiveness factor values of D-amino acid oxidase immobilized in calcium pectate gel studied by flow microcalorimetry [28]

Parameters	Cell loading concentration (mg dry mass mL^{-1} gelling suspension)				
	24.0	34.2	48.0	60.0	79.9
k_1 (min^{-1})	0.45	0.60	1.45	1.18	4.50
Thiele modulus, Φ (–)	3.08	3.55	5.52	4.98	9.73
Effectiveness factor, η (–)	0.66	0.60	0.44	0.47	0.27

loading on the activity and effectiveness of the biocatalyst. The advantage of this approach is the ease and rapidity with which the experimental data are treated. This is very useful for a simple comparison of the intrinsic activity of different biocatalysts.

In spite of usefulness of the simplification obtained by decreasing the experimental substrate concentration, many studies are aimed at the investigation of kinetic properties of immobilized biocatalysts within broader concentration ranges. In a previous paper [29], cells of *Escherichia coli* with penicillin acylase activity were immobilized by entrapment in calcium pectate gel and tested on the transformation of penicillin G to 6-amino penicillanic acid. Figure 9 shows experimental data from a microcalorimetric investigation of the penicillin G transformation in steady state. As appreciable particle-mass transfer was expected, the mathematical model that includes particle-mass balance was used.

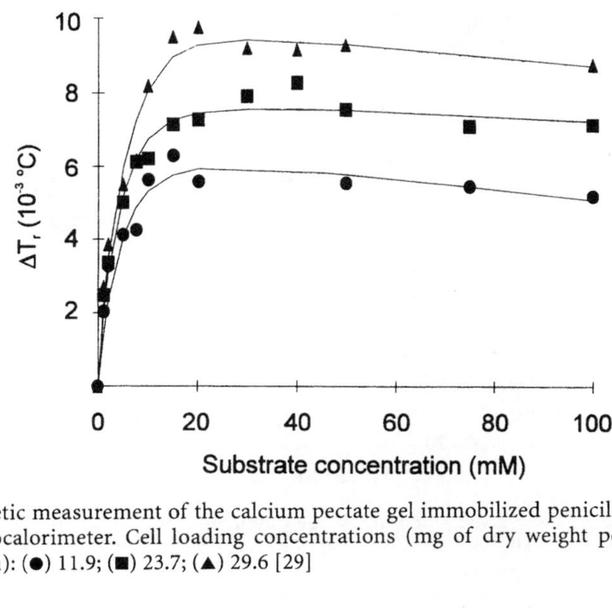

Fig. 9. Kinetic measurement of the calcium pectate gel immobilized penicillin acylase in the flow microcalorimeter. Cell loading concentrations (mg of dry weight per mL of gelling suspension): (●) 11.9; (■) 23.7; (▲) 29.6 [29]

The experimental data shown in Fig. 9 were used for the determination of the kinetic parameters. The kinetic model of substrate inhibition was in the form:

$$v(c_S, c_P, \mathbf{P}) = \frac{V_m c_{Sb}}{K_m + c_{Sb} + \dfrac{c_{Sb}^2}{K_i}} \tag{35}$$

$$v_{obs} = \eta \frac{1-\varepsilon}{\varepsilon} \frac{V_m c_{Sb}}{K_m + c_{Sb} + \dfrac{c_{Sb}^2}{K_i}} . \tag{36}$$

The thermometric data in Fig. 9 were transformed via parameter α, which was determined by calibration [29] and used to determine the kinetic parameters V_m, K_m and K_i in Eq. (36) by nonlinear regression. The effectiveness factor values were calculated by solving Eq. (7) using the orthogonal collocation procedure [37]. As can be seen in Fig. 9, the kinetic behavior corresponds to a reaction slightly inhibited by substrate. When the bulk concentration of the substrate exceeds the maximum on the kinetic curve, the resistance to mass transfer can increase the reaction rate by decreasing the actual substrate concentration in the enzyme site. Generally, this situation is expressed by effectiveness factor values being greater than 1. Values of the effectiveness factor calculated on the basis of kinetic parameters obtained for penicillin acylase are represented in Fig. 10.

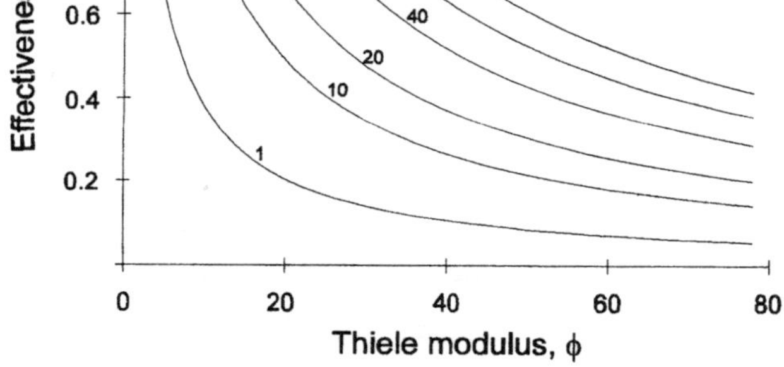

Fig. 10. Effectiveness factor values for immobilized penicillin acylase calculated by Eq. (6). Kinetic parameter values ($K_m = 1.5$ mM, $K_i = 570$ mM) were calculated from results in Fig. 11. Parameter of lines is input penicillin G concentration in mM [29]

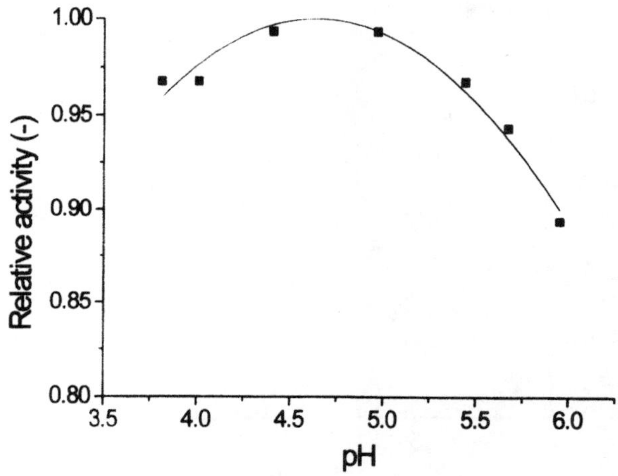

Fig. 11. Microcalorimetric investigation of pH-activity profile of invertase immobilized on Eupergit C by direct binding (unpublished results)

5.2
pH-Activity Profiles

The investigation of pH-activity profiles is a typical measurement when a relative activity change provides sufficient information about the enzyme properties. Equation (21) reminds us that the actual reaction rate in the column is directly proportional to the measured thermometric signal, on condition that the column is packed with a differential amount of IMB. Since the reaction rate is proportional to the IMB activity, the activity is proportional to the thermometric signal, as well. Therefore, the measurement of the relative change of thermometric signal due to the pH change can be used instead of conventional techniques for the investigation of pH-activity. Once more, enzyme flow microcalorimetry proves to be a robust, efficient and accurate method.

An example of such a result is illustrated in Fig. 11 showing the pH-activity profile of invertase immobilized on Eupergit C. The relative activity plotted in the figure is the ratio of the thermometric signal at given pH divided by the maximum value of the thermometric signal observed at the pH optimum. The value of the pH optimum obtained is comparable to the known value for yeast invertase [38].

5.3
Biocatalyst Stability

As in the case of pH-activity measurement, the investigation of enzyme stability is quite often satisfied by obtaining information about the relative decrease of activity during the enzyme operation, without knowing the absolute value of the activity during the experiment. Figure 12 shows the stability of invertase bound

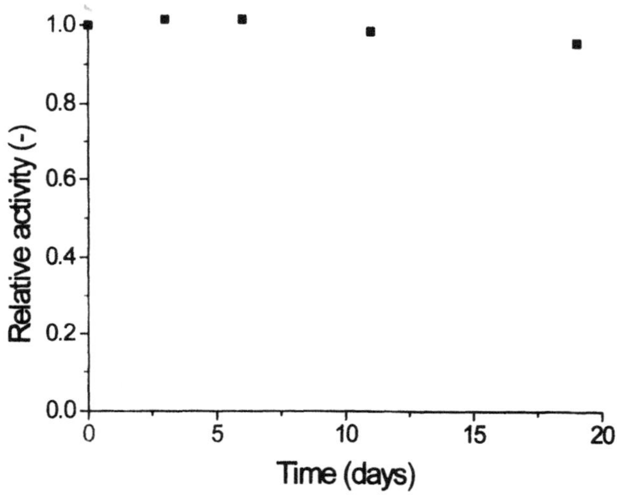

Fig. 12. Operational stability of invertase immobilized on Eupergit C investigated by flow enzyme microcalorimetry. Temperature 30 °C, flow rate 1 mL min^{-1} unpublished results

on Eupergit C and measured by flow microcalorimetry. The thermometric data were normalized using the initial value of thermometric signal as a norm.

The operational stability of *Trigonopsis variabilis* cells with D-amino acid oxidase activity, entrapped in standard and hardened ionotropic gels, was also investigated by means of the FMC [39]. The activity of the biocatalyst packed in the FMC column was continuously monitored by the FMC signal measurement

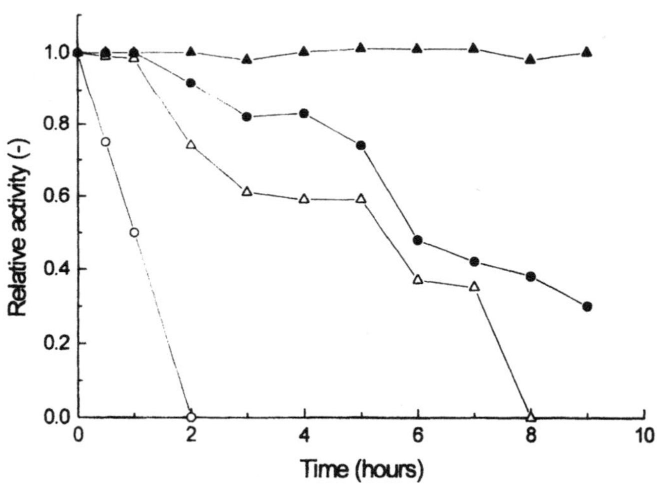

Fig. 13. Operational stability of D-amino acid oxidase fixed in cells of *Trigonopsis variabilis* CCY 15-1-3 entrapped in standard (△) and hardened (▲) calcium pectate gel and standard (○) and hardened calcium alginate gel (●). The relative activity was monitored by continuous processing, with the substrate (cephalosporin C) solution in the flow microcalorimeter [39]

when cephalosporin C solution was fed into the system. The results shown in Fig. 13 clearly demonstrate the superior stability of hardened calcium pectate gel biocatalyst particles.

Another type of stability of immobilized biocatalysts is the retention of activity after periodic use in batch processes, as has been reported previously for penicillin acylase entrapped in polyacrylamide gel [40]. This option can be used to advantage for rapid monitoring of biocatalyst activity under conditions of industrial application. Apart from the measurement of activity as an indication of the necessity to replace the biocatalyst, the periodic analysis of the variation of kinetic properties permits greater insight into deviation from the optimal parameters.

Flow microcalorimetry allows the investigation of the relation between environmental conditions and enzyme stability. For example, the influence of ethanol concentration on the stability of immobilized invertase was studied this way [33].

5.4 Inhibition

Flow microcalorimetry was used for the investigation of product inhibition in the urea–urease system [25]. Inhibition was induced by the product of the reaction. As mentioned above, this was a case in which the model had to be in the form of differential equations, and the mass and heat balance had to be solved by integration of these equations along the column. Another approach was adopted in the investigation of invertase inhibition by products of sucrose hydrolysis [33]. In this work, inhibition was observed qualitatively when samples of sucrose with different concentrations of products (glucose or fructose) were fed into the immobilized enzyme column of the microcalorimeter. Unfortunately, there are only a few reports nowadays about enzyme inhibition analysis by flow microcalorimetry at the level of mathematical modeling published in the literature. The majority of studies are concerned with the inhibitory effect of substrate [27, 29–31]. In these investigations the mathematical treatment of data is presented in simplified form, using the approaches described above. Many kinetic models, including the inhibitory effects of compounds on an enzyme, fit a kinetic equation of the following general form

$$v(c_{Sb}, c_{Pb}, \mathbf{P}) = \frac{V_m c_S}{f(K_m, K_i, c_S, c_i)} \tag{37}$$

where c_i is the inhibitor concentration and the function f in the denominator is a general representation of denominators in kinetic equations involving inhibition. Thus, when thermometric data are converted to reaction rates, the following equation can be used for their description

$$v_{obs} = \eta \frac{1-\varepsilon}{\varepsilon} \frac{V_m c_S}{f(K_m, K_i, c_S, c_i)} \,. \tag{38}$$

The analysis of substrate inhibition based on Eq. (38) was performed for immobilized penicillin acylase as mentioned above. However, if particle mass transfer can be neglected, an even simpler approach can be adopted, using the equation that is valid for the differential bed

$$\Delta T_r = \left(\alpha \frac{1-\varepsilon}{\varepsilon} V_m\right) \frac{c_S}{f(K_m, K_i, c_S, c_i)} = \frac{\alpha' c_S}{f(K_m, K_i, c_S, c_i)} \, . \tag{39}$$

Since quantities in the first parenthesis are constant, they can be regrouped in a single parameter, α'. Thus, all parameters in the denominator of the kinetic term can be determined directly by nonlinear regression from thermometric data. Combining Eq. (39) with the kinetic model of the sucrose hydrolysis catalyzed by invertase represented by Eq. (25) we obtain

$$\Delta T_r = \frac{\alpha' c_{Sb}}{K_m + c_{Sb} + \frac{c_{Sb}^2}{K_i}} \frac{W}{W_0} \, . \tag{40}$$

The last equation was used in several papers studying the kinetic properties of invertase, where substrate inhibition was evaluated in terms of the inhibition parameter K_i [27, 30, 31].

5.5 Screening of Immobilized Biocatalysts

Flow microcalorimetry, which makes many rapid and accurate measurements of the activity of immobilized biocatalysts, provides a tool for researchers that can be used to discriminate between different preparatives of immobilized biocatalysts. Table 3 shows previous experiments where the characterization of kinetic properties by flow microcalorimetry was used to compare different techniques of purified enzyme immobilization [27, 30, 31, 35] as well as the immobilization of enzymes fixed in cells [28, 29, 40]. More details can be found in our recent review article [41].

5.6 Monitoring of Enzymes Immobilized on Affinity Sorbents

While the interest in irreversible and covalent methods of immobilization continues, bioaffinity-based procedures are gaining appreciable attention, especially for analytical applications [42]. Compared to other methods of immobilization, those based on bioaffinity offer several distinct advantages. These include:

1. Binding of the enzyme to an affinity support may be very strong and yet reversible under specific conditions;
2. The immobilization process is usually simple, and necessitates no special skills;

Table 3. Screening of immobilized biocatalysts by means of the flow microcalorimetry

Biocatalyst	Form	Origin	Mode of immobilization	Reference
Invertase	Enzyme	*S. cerevisiae*	Covalent	[27]
	Enzyme	*S. cerevisiae*	Biospecific	[30]
	Enzyme	*S. cerevisiae*	Biospecific + covalent	[31]
	Enzyme	*S. cerevisiae*	Biospecific	[35]
D-Amino acid oxidase	Enzyme	*T. variabilis* [a]	*Cross*-linking	[28]
	Cells	*T. variabilis*	Entrapment	[28]
	Cells	*T. variabilis*	Entrapment + *cross*-linking	[28]
Penicillin acylase	Enzyme	*E. coli*	Entrapment + *cross*-linking	[40]
	Cells	*E. coli*	Entrapment + *cross*-linking	[40]
	Cells	*E. coli*	Entrapment	[29]

[a] *Trigonopsis variabilis.*

3. The support matrix is reusable;
4. Oriented immobilization facilitates good expression of activity and stabilization against inactivation;
5. Direct immobilization of the enzyme from partially pure preparations is possible.

As an example, we have chosen the model "bioaffinity-immobilization of glycoenzymes on lectins, precoupled on water-insoluble matrices". The largest and best characterized is concanavalin A (Con A), that belongs to the legume family. Con A has been widely used in the affinity purification of a variety of glycoenzymes and nearly exclusively in the immobilization of glycoenzymes on a variety of supports. A good review of the use of Con A in glycoenzyme immobilization has recently been published [43].

The binding of the glycoenzyme to the Con A-precoupled affinity support may be very strong, but it is nevertheless reversible under specific conditions. A biospecific adsorption of invertase to pre-coupled Con A is the best example. Con A-immobilized invertase in the FMC packed bed reactor retained full operational stability for several hours/days of processing in the flow, even at high substrate concentration [31]. In spite of the extraordinarily strong binding of invertase to the Con A-bead cellulose conjugate ($K_D = 5 \times 10^{-9}$ M), conditions have been found for the use of a Con A column as an affinity chromatography medium. The determining factor in the release of bound invertase by a counterligand (α-methyl-D-*manno*pyranoside) is the time of incubation. This phenomenon was demonstrated in both batch and flow-through experiments by the FMC. It was also shown that the ease of elution or optimal recovery of the enzyme depends upon the lectin concentration on the support [44].

A method based on the FMC determination of catalytic activity of IMB in the FMC column was used to monitor the process of lectin affinity chromatography of invertase on Con A-bead cellulose [45]. Adsorption and desorption in the chromatography column were examined by FMC in small samples withdrawn from the column. Attention was given to the operating parameters and the storage stability of the affinity sorbent. The binding ability of the affinity matrix decreased with the number of consecutive chromatographic runs, although its storage stability was satisfactory.

Binding of Con A to glycoproteins could be prevented/reversed by several sugars which are known to interact with the lectin, such as α-methyl-D-*gluco*-pyranoside, D-mannose or α-methyl-D-*manno*pyranoside [46]. Glycosylated enzymes, in this case invertase and glucose oxidase, could be used as competitive markers for the simple and rapid determination of the relative carbohydrate specificity of lectins for mono- and oligosaccharides. They were used as the competitive saccharides in homogeneous and heterogeneous systems: (lectin + enzyme + saccharide) and (lectin immobilized on support + enzyme + saccharide), respectively, under static or flow conditions. The activity of the enzyme was determined indirectly by spectrophotometry in the supernatant liquid, after the precipitated lectin-enzyme complex was separated by centrifugation (homogeneous system) or in a Con A-sorbent with specifically bound enzyme (heterogeneous system at static conditions). A new, alternative method using enzyme flow microcalorimetry was applied in flow conditions for the direct determination of the biospecifically adsorbed enzyme. The relative carbohydrate specificity of lectins (Con A) was estimated as the concentration of the saccharide (D-mannose, α-methyl-D-*manno*pyranoside) that inhibited the interaction to 50%. The results obtained by a precipitation-inhibition assay in solution and a sorption-inhibition assay in a batch system intercorrelated significantly with those obtained with a sorption-inhibition assay in the FMC flow system [46].

Among generally appropriate ligands, specific antienzyme antibodies are clearly the most versatile, being applicable – at least theoretically – to virtually every enzyme. Moreover, anti-Con A antibody may also be employed, at least in an analytical determination of freely accessible Con A pre-linked to affinity sorbent. The amounts of accessible Con A pre-linked to bead cellulose could be evaluated: (1) by adsorption of invertase, and (2) by means of ELISA, using both mouse anti-Con A antibody and pig anti-mouse antibody linked with peroxidase [35]. Reactions under conditions 1 and 2 were monitored in two ways: thermometrically, by FMC, as well as by conventional spectrophotometry. The results obtained by spectrophotometry and thermometry correlated significantly ($p = 0.028$ and $p = 0.0001$, for ELISA and the invertase). The amounts of immobilized Con A were directly proportional to the ELISA signal of the peroxidase reaction as detected by FMC ($r = 0.983$, $p = 0.0005$). A slightly less tight correlation was found between the amounts of immobilized Con A and the activity of invertase in conjugates when they were assayed by means of thermometry ($r = 0.933$, $p = 0.0065$) and spectrophotometry ($r = 0.920$, $p = 0.0094$).

These results indicate that flow microcalorimetry may be considered to be a method for rapid evaluation of the amount and accessibility of immobilized Con A [35].

The proposed method of IME screening assisted by the FMC has significantly simplified research directed towards increasing the amount of Con A and glycoenzyme associated with solid supports. A simple strategy that has recently been developed by the authors involves building up layers of Con A and glycoenzymes on bead cellulose [31, 47] and poly(glycidyl methacrylate) (Eupergit C) [47] matrices precoupled with Con A. Bioaffinity-layered preparations of invertase exhibited – layer by layer – up to a tenfold increase in catalytic activity, as measured by FMC [31,47]. A somewhat similar approach was later used by Farooqui et al. [48] in the preparation of superactive immobilized glucose oxidase, improving the sensitivity of glucose oxidase-based glucose monitoring FET-biosensor devices. It is anticipated that the technique will be applicable to all of the glycoenzymes investigated, namely glucose oxidase, invertase, β-galactosidase and amyloglucosidase [42] for both FMC and FET biosensors. Tailor-made neoglycoenzymes [49], exhibiting a similar effect of *bioaffinity layering*, should be considered for FMC signal amplification, as well.

6
General Strategy for FMC Experiments

Current knowledge, summarized in this review, is sufficient to define a general strategy for the study of immobilized biocatalyst properties by flow microcalorimetry. The principal simplification of the investigation of the kinetics of immobilized biocatalysts consists in using only a small amount of the IMB and working with low substrate conversions. The FMC column can then be considered to be a differential reactor and the mathematical description is reduced to the system of algebraic instead of differential equations. Another important improvement is the developed procedure of "autocalibration" of the FMC, leading to the value of parameter α, enabling the determination of reaction rates from thermometric data. In this way, the thermometric data are converted to more conventional, mass-balance data, which can be treated by the mathematical tools that are generally used for calculating the rates of reactions catalyzed by immobilized biocatalysts.

The optimal organization of the kinetic FMC experiment is proposed in the flowchart depicted in Fig. 14. The procedure can be divided into the following principal points:

1. Experiments in the flow-through mode, in which the FMC operates in steady-state mode. Here, first of all, the appropriate amount of IMB in the column to be used is determined. Then, the steady state kinetic data – the dependence of the thermometric signal on substrate concentration – is measured;
2. Experiments in the total recirculation mode, in which the FMC operates as a batch system in non-steady-state. The dependence of the thermometric signal on time for a chosen substrate concentration is measured;
3. The mathematical treatment of the data depends on whether the reaction is diffusion limited or not. This part of work depends on the aim of the study. If

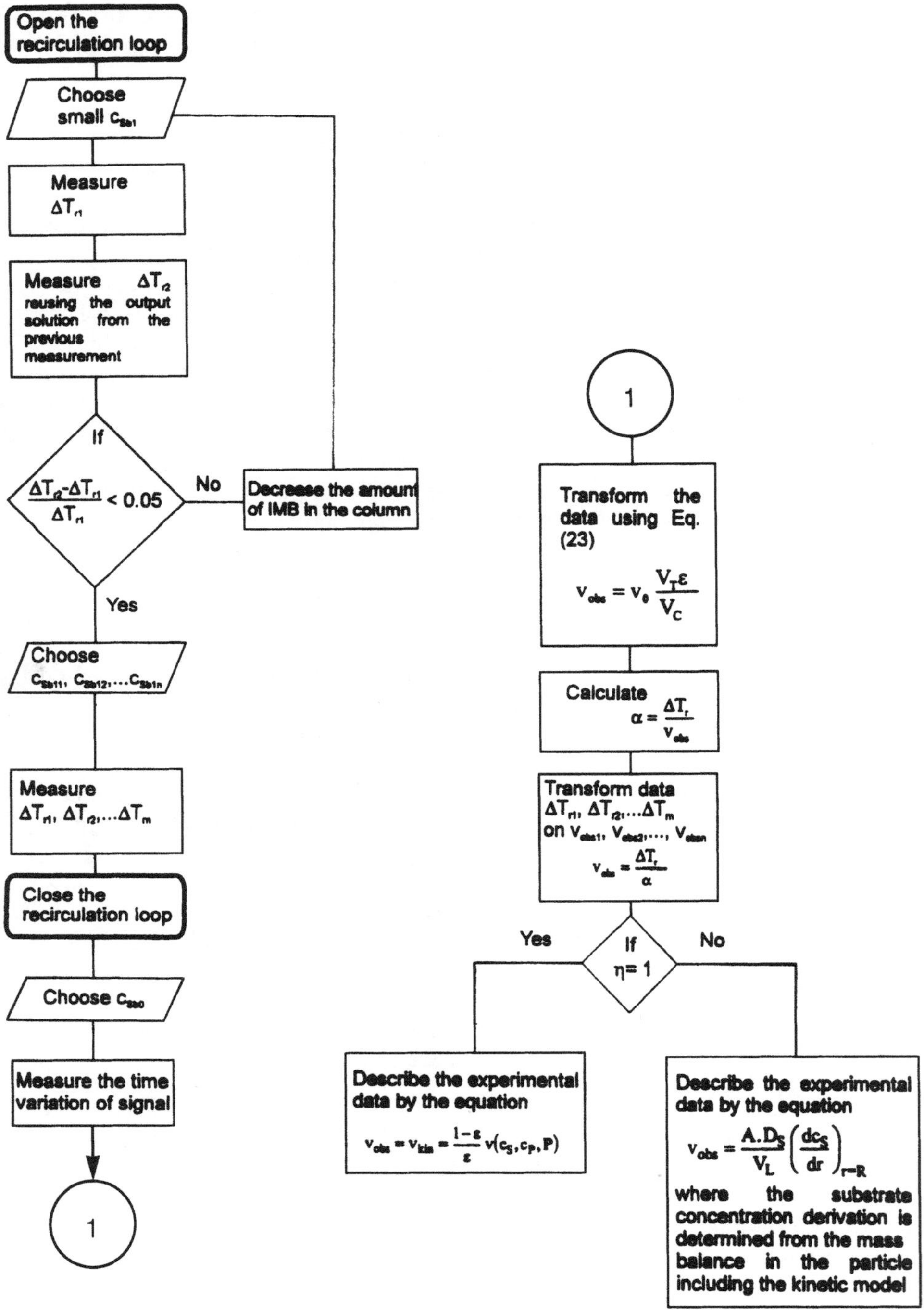

Fig. 14. Optimal procedure for the kinetic characterization of immobilized biocatalysts by flow microcalorimetry

the reaction is not diffusion limited, the thermometric data are quite often used directly for the determination of kinetic parameters in the denominator of the kinetic equation (e.g. K_m, K_i). Similarly, the thermometric data provide information for direct comparison of the relative activity of different IMB preparations, or of the same IMB under varied reaction conditions (e.g. pH-activity profile). On the other hand, for the treatment of data obtained for a diffusion-limited reaction, or when information about absolute biocatalyst activity is required, it is convenient to convert the thermometric data to reaction rate values. In order to do this, the transformation parameter, α, is calculated from the experiment in the infinite recirculation mode. Then, after the data transformation, further computation is based on material balance equations.

7
Concluding Remarks

Experience shows that flow microcalorimetry is a universal technique that is suitable for the investigation of the catalytic properties of immobilized biocatalysts. This review has summarized all basic examples of its application, but has not exhausted all of their potential possibilities. As an example, the steady-state measurement of a bi-substrate enzyme reaction with a *co*-immobilized glucose oxidase-catalase system was reported [26]. However, there is no report on the evaluation of kinetic properties of partial enzymes in *co*-immobilized systems. Even the measurement of the overall heat produced in such systems does not provide direct information about partial reactions. We believe that new approaches to analyze these systems based on mathematical modeling can be developed.

Besides steady state measurements, there is probably good reason to use flow microcalorimetry for the study of non-steady-state behavior in systems with immobilized biocatalysts. Here, the mathematical description is more complicated, requiring the solution of partial differential equations. Moreover, the heat response can evolve non-specific heats, like heat of adsorption/desorption or mixing phenomena. In spite of these complications, the possibility of the on-line monitoring of the enzyme reaction rate can provide a powerful tool for studying the dynamics of immobilized biocatalyst systems.

Finally, the methodology summarized in this review indicates how easy it would be to design universal measuring systems capable of performing experiments on the catalytic properties of enzymes in an automatic way. This too is our aim.

8
References

1. The Working Party on Immobilized Biocatalysts within the European Federation of Biotechnology (1983) Enzyme Microb Technol 5:304
2. Ford JR, Lambert AH, Cohen W, Chambers RP (1972) Biotechnol Bioeng Symp 3:267
3. Buchholz K, Klein J (1987) Characterization of immobilized biocatalysts. In: Mosbach K (ed) Methods in Enzymology, vol. 135. Immobilized Enzymes and Cells, Part B. Academic Press, Orlando, p 3

4. Carr PW, Bowers LD (1980) In: Immobilized Enzymes in Analytical and Clinical Chemistry Fundamentals and Applications. John Wiley, New York, p 455
5. Johansson A, Lundberg J, Mattiasson B, Mosbach K (1973) Biochim Biophys Acta 304:217
6. Cooney CL, Weaver JC, Tannenbaum SR, Faller DV, Shieds A, Janke M (1974) In: Pye EK, Wingard LB (eds) Enzyme Engineering, vol. 2. Plenum Press, New York, p 411
7. Bowers LD, Carr PW, Schifreen RS (1976) Clin Chem 22:1427
8. Mosbach K, Danielsson B (1974) Biochim Biophys Acta 364:140
9. Canning LM, Carr PW (1975) Anal Lett 8:359
10. Mosbach K, Danielsson B, Bogerud A, Scott M (1975) Biochim Biophys Acta 403:256
11. Schmidt HL, Krisam G, Grenner C (1976) Biochim Biophys Acta 429:283
12. Krisam GZ (1978) Anal Chem 280:130
13. Bowers LD, Canning LM, Sayers KM, Carr PW (1976) Clin Chim Acta 22:1314
14. Spink C, Wadsö I (1976) Methods Biochem Anal 23:1
15. Danielsson B, Mosbach K (1988) Enzyme thermistors. In: Mosbach K (ed) Methods in Enzymology, vol 137. Immobilized Enzymes and Cells, Part B. Academic Press, Orlando, p 181
16. Danielsson B (1979) The Enzyme Thermistor. Development and Application. Thesis, Department of Biochemistry, Chemical Center, University of Lund, Lund
17. Decristoforo G (1988) Flow-injection analysis for automated determination of beta-lactams using immobilized enzyme reactors with thermistor or ultraviolet spectrophotometric detection. In: Mosbach K (ed) Methods in Enzymology, vol 137. Immobilized Enzymes and Cells, Part D. Academic Press, Orlando, p 197
18. Danielsson B, Mosbach K (1987) Enzyme Thermistors. In: Turner APF, Karube I, Wilson GS (eds) Biosensors: Fundamentals and Applications. Oxford University Press, Oxford, p 575
19. Kirstein D, Danielsson B, Scheller F, Mosbach K (1989) Biosensors 4:231
20. Satoh I (1988) Biomedical Applications of the Enzyme Thermistor in Lipid Determinations. In: Mosbach K (ed) Methods in Enzymology, vol 137. Immobilized Enzymes and Cells, Part D. Academic Press, Orlando, p 217
21. Birnbaum S, Bülow L, Danielsson B, Mosbach K (1988) Automated TELISA. In: Mosbach K (ed) Methods in Enzymology, vol 137. Immobilized Enzymes and Cells, Part D. Academic Press, Orlando, p 343
22. Mandenius CF, Danielsson B (1988) Enzyme Thermistors for Process Monitoring and Control. In: Mosbach K (ed) Methods in Enzymology, vol 137. Immobilized Enzymes and Cells, Part D. Academic Press, Orlando, p 307
23. Eftink MR, Johnson RE, Biltonen RL (1981) Anal Biochem 11:305
24. Beezer AE, Steenson TI, Tyrrel HJV (1974) Talanta 21:467
25. Owusu RK, Trewhella MJ, Finch A (1985) Biochim Biophys Acta 830:282
26. Owusu RK, Finch A (1986) Biochim Biophys Acta 872:83
27. Štefuca V, Gemeiner P, Kurillová L', Danielsson B, Báleš V (1990) Enzyme Microb Technol 12:830
28. Gemeiner P, Štefuca V, Welwardová A, Michálková E, Welward L, Kurillová L', Danielsson B (1993) Enzyme Microb Technol 15:50
29. Štefuca V, Welwardová A, Gemeiner P, Jakubová A (1994) Biotechnol Tech 8:497
30. Dočolomanský P, Gemeiner P, Mislovičová D, Štefuca V, Danielsson B (1994) Biotechnol Bioeng 43:286
31. Gemeiner P, Dočolomanský P, Nahálka J, Štefuca V, Danielsson B (1996) Biotechnol Bioeng 46:26
32. Štefuca V, Vikartovská-Welwardová A, Gemeiner P (1997) Anal Chim Acta 355:63
33. Maske M, Strauss A, Kirstein D (1993) Anal Lett 26:1613
34. Bowski L, Saini R, Ryu DY, Vieth W (1971) Biotechnol Bioeng 13:641
35. Dočolomanský P, Breier A, Gemeiner P, Ziegelhöffer A (1995) Anal Lett 28:2585
36. Satterfield CN (1976) Mass transfer in heterogeneous catalysis, Russian translation. Chimija, Moskva
37. Villadsen J, Stewart WE (1967) Chem Eng Sci 22:1483

38. Lampen JO Boyer PD (1971) In: Boyer PD (ed) The Enzymes, Vol. V: Hydrolysis, Hydration. Academic Press, New York, pp 291-305
39. Gemeiner P, Nahálka J, Vikartovská A, Nahálková J, Tomáška M, Šturdík E, Markovič O, Malovíková A, Zatková I, Ilavský M (1996) In: Wijffels RH, Buitelaar RM, Bucke C, Tramper J (eds) Progress in Biotechnology, vol 11. Immobilized Cells: Basics and Applications, Elsevier, Amsterdam, pp 76-83
40. Welwardová A, Gemeiner P, Michálková E, Welward L, Jakubová A (1993) Biotechnol Tech 7:809
41. Gemeiner P, Štefuca V, Danielsson B (1996) In: Bittan EE, Danielsson B, Bülow L (eds) Advances in Molecular and Cell Biology, Vol. 15 B. Biochemical Technology, JAI Press Inc, Greenwich, pp 411-418
42. Saleemuddin M (1998) In: Scheper T (ed) Advances in Biochemical Engineering/Biotechnology. Springer, Berlin Heidelberg New York (in press)
43. Saleemuddin M, Husain Q (1991) Enzyme Microb Technol 13:290
44. Mislovičová D, Chudinová M, Gemeiner P, Dočolomanský P (1995) J Chromatogr B 664:145
45. Mislovičová D, Chudinová M, Vikartovská A, Gemeiner P (1996) J Chromatogr A 722:143
46. Mislovičová D, Vikartovská A, Gemeiner P (1997) J Biochem Biophys Methods 35:37
47. Gemeiner P, Dočolomanský P, Vikartovská A, Štefuca V (1998) Biotechnol Appl Biochem 22, 000-000, in press
48. Farooquí M, Saleemuddin M, Ulber R, Sosnitza P, Scheper T (1997) J Biotechnol 55:171-179
49. Masárová J, Mislovičová D, Gemeiner P (1998) XIVth Biochemical Meeting with international participation, Stará Lesná, October 12-15, 1998; contribution will be published as short communication in Chem Papers

Bioactive Agents from Natural Sources: Trends in Discovery and Application

Susanne Grabley[1] · Ralf Thiericke[2]

Hans-Knöll-Institut für Naturstoff-Forschung e.V., Beutenbergstrasse 11, D-07745 Jena, Germany

[1] *E-mail: sgrabley@pmail.hki-jena.de*
[2] *E-mail: thierick@pmail.hki-jena.de*

About 30% of the worldwide sales of drugs are based on natural products. Though recombinant proteins and peptides account for increasing sales rates, the superiority of low-molecular mass compounds in human diseases therapy remains undisputed mainly due to more favorable compliance and bioavailability properties. In the past, new therapeutic approaches often derived from natural products. Numerous examples from medicine impressively demonstrate the innovative potential of natural compounds and their impact on progress in drug discovery and development. However, natural products are currently undergoing a phase of reduced attention in drug discovery because of the enormous effort which is necessary to isolate the active principles and to elucidate their structures. To meet the demand of several hundred thousands of test samples that have to be submitted to high-throughput screening (HTS) new strategies in natural product chemistry are necessary in order to compete successfully with combinatorial chemistry. Today, pharmaceutical companies have to spend approximately US $350 million to develop a new drug. Currently, approaches to improve and accelerate the joint drug discovery and development process are expected to arise mainly from innovation in drug target elucidation and lead finding. Breakthroughs in molecular biology, cell biology, and genetic engineering in the 1980 s gave access to understanding diseases on the molecular or on the gene level. Subsequently, constructing novel target directed screening assay systems of promising therapeutic significance, automation, and miniaturization resulted in HTS approaches changing the industrial drug discovery process drastically. Furthermore, elucidation of the human genome will provide access to a dramatically increased number of new potential drug targets that have to be evaluated for drug discovery. HTS enables the testing of an increasing number of samples. Therefore, new concepts to generate large compound collections with improved structural diversity are desirable.

Keywords: Automation, Biological derivatization, Chemical screening, Combinatorial natural product chemistry, Drug discovery, High-throughput screening, Natural products, Physico-chemical screening, Screening bioassays, Structural diversity from nature, Target identification.

Advances in Biochemical Engineering/
Biotechnology, Vol. 64
Managing Editor: Th. Scheper

List of Abbreviations

BMS	Bristol Myers Squibb
CAMD	computer aided molecular design
cDNA	complementary desoxyribonucleic acid
CE	capillary electrophoresis
cGMP	cyclic guanosine monophosphate
DAD	diode-array detection
DMSO	dimethyl sulfoxide

DNA	desoxyribonucleic acid
ELISA	enzyme linked immunosorbent assay
ESI	electrospray ionisation
ESTs	expressed sequence tags
FCS	fluorescence correlation spectroscopy
FDA	Food and Drug Administration
FKBP	FK506 binding protein
GABA	*gamma* aminobutyric acid
GBF	Gesellschaft für Biotechnologische Forschung (Braunschweig; Germany)
HGS	Human Genome Sciences
HIV	Human immunodeficiency virus
HKI	Hans-Knöll-Institut for Natural Product Research (Jena, Germany)
HMG-CoA	3-hydroxy-3-methylglutaryl coenzyme A
HPLC	high performance (pressure) liquid chromatography
HPLC-ESI-MS	high performance (pressure) liquid chromatography electrospray mass spectrometry
HPTLC	high performance thin-layer chromatography
HTS	high-throughput screening
IC	inhibitory concentration
IR	infra-red
LD	lethal dosis
LC-MS	liquid chromatography-mass spectrometry
LC-NMR	liquid chromatography-nuclear magnetic resonance
mRNA	messenger ribonucleic acid
MS	mass spectrometry
NADP	nicotinamide-adenine dinucleotide phosphate
NCE	new chemical entity
NCI	National Cancer Institute
NMR	nuclear magnetic resonance
PCR	polymerase chain reaction
Rhône-PR	Rhône-Poulenc Rorer
RNA	ribonucleic acid
RP-HPLC	reversed phase high performance (pressure) liquid chromatography
SAR	structure activity relationship
SKB	SmithKline Beecham
SPA	scintillation proximity assay
SPE	solid phase extraction
TPA	tetradecanoylphorbol-13-acetate
TIGR	The Institute for Genomic Research
TLC	thin-layer chromatography
TLC-FID	thin-layer chromatography flame ionization detection
UHTS	ultra high-throughput screening
UV	ultraviolet
UV/VIS	ultraviolet/visuable

1 Introduction

1.1 Natural Products in Drug Discovery and Development

Although interest in natural products as a source of innovation in drug discovery has decreased in the last few decades, therapeutics of microbial or plant origin count for more than 30% of the current worldwide human therapeutics sales. Considering the top-twenty drugs of 1997 approximately 40% derive from nature, and in the time period from 1992 to 1996 34% of the new chemical entities (NCE) introduced were natural compounds. For anticancer and anti-infective treatment even 60% of the approved drugs and the pre-new drug application candidates (excluding biologics) are of natural origin for the period 1989 to 1995 [1]. These figures demonstrate the evidence of natural product research, and its impact on drug discovery and development. Furthermore, data reflecting recent trends emphasize the superior role of natural products as both, valuable lead compounds and potential new drugs [2].

In the future, novel chemical structures with improved therapeutic properties by affecting new disease targets can be expected to arise from genomic sciences. Structural and functional analysis of the human genome gives access to an increasing number of disease related genes and their products, thus, hopefully contributing to more causal therapies [3]. In order to drive foreward the drug discovery and development process an active principle substantially affecting a disease target is requested. In most cases, first hits are discovered by random screening approaches. Subsequent validation steps follow to answer the key question whether a hit will give rise to a lead compound that will prove beneficial for optimizing structure/activity relationship by the well established powerful tools of rational drug design. Issues of interest not only comprise bioactivity at the target of interest but also applicability, bioavailability, biostability, metabolism, toxicity, specificity of bioactivity, distribution, tissue selectivity, cell penetration properties, and as a key prerequisite for development, access to large quantities of the compound for clinical studies and commercialization has to be guaranteed [4].

In the past, compounds from nature often opened up completely new therapeutic approaches. Moreover, natural compounds substantially contributed to identify and understand novel biochemical pathways in vitro and in vivo, and consequently proved to make available not only valuable drugs but also essential tools in biochemistry, and molecular cell biology. It is worth while studying exhaustively the molecular basis of biological phenomena of new and/or unusual chemical structures from nature. Numerous examples from medicine impressively demonstrate the innovative potential of natural products and their impact on progress in drug discovery and development as discussed below.

Further examples can be added from agricultural use:

i) The herbicide phosphinothricin (glufosinate, Basta) represents a unique methyl phosphinic acid analog of glutamic acid acting as a strong competitive inhibitor of glutamine synthetase bearing only low toxicity [5]. The antibacterial tripeptide L-phosphinothricyl-L-alanyl-L-alanine (bialaphos) can be isolated from culture broths of *Streptomyces viridochromogenes* [6], and *Streptomyces hygroscopicus* [7]. Antibiotic resistance of the producing organism is due to enzymatic *N*-acetylation of a biosynthetic tripeptide precursor [8]. To broaden the application range of phosphinothricin as a herbicide the gene for the corresponding *N*-acetyltransferase was successfully transferred to plants in order to generate Basta resistant transgenic crops [9]. Basta was commercialized by the German company Hoechst as a racemate derived from chemical synthesis.

ii) The avermectins isolated from *Streptomyces avermitilis* exhibit broad insecticidal, ascaricidal, and antihelmintic activity even at very low concentrations without showing any antibacterial or antifungal activity. They were discovered by using an animal model to detect anticoccidial agents [10]. Detailed studies proved that the avermectins interact in a unique manner with the GABA regulated chloride ion channels. Chemical modification of the macrolide type avermectins resulted in ivermectin which was commercialized as a mixture of ivermectin B_{1a} and B_{1b} by the US company Merck for application in agriculture and animal health care, as well as in human medicine to prevent river blindness in the developing countries of Africa and Central America [11].

iii) Most recently, representatives of a completely new class of non-phytotoxic agro-fungicides were successfully brought onto the market by the German company BASF and the English company ICI. The compounds were developed by total synthesis starting from the fungal metabolites strobilurin and oudemansin discovered by T. Anke and W. Steglich as lead structures [12, 13]. Antifungal activity was proven to be due to the inhibition of the mitochondrial bc_1 complex of the fungal respiratorial chain. Toxicity is prevented by rapid metabolic processes in mammals and numerous other eukaryotic organisms.

1.2 High-Throughput Screening Towards Lead Discovery

Breakthroughs in molecular biology, cell biology, and genetic engineering in the 1980s gave rise to understanding diseases on the molecular or on the gene level. Subsequently, constructing novel target directed screening systems of promising therapeutic significance resulted in changing industrial drug discovery attempts completely [3, 4, 14]. Thus, during the last few years a large number of different molecular screening assays has been developed. Transfer to the 96-well or, more recently, to the 384-well microplate format, automation and miniaturization turned out to be crucial with respect to industrial high-throughput screening (HTS). Down-scaling of the assay volume to the 1,536-well format is already underway, although considerable problems have to be solved mainly concerning liquid-handling in the nanoliter range and assay read-out. However, approaches towards the testing of 100,000 samples per day are in progress.

Lead discovery strategies for both, pharmaceutical, and agrochemical applications are in a highly rapid developmental stage. The driving force is the reduction in the time to reach the market for new drugs which requires dramatic performance improvements in lead finding and development. Success and failure are tremendously coupled to both, the novelty and meaningfulness of the applied biological test systems, and the number and structural diversity of test compounds available. Unambiguous infrastructural elements for effective HTS are target discovery and evaluation, screen design, sample sourcing, robotic hardware, and data management. Therefore, laboratory automation is an indispensible tool for the efficiency in HTS, and plays a crucial role in the early lead-finding stages of drug discovery. The deployment of automation bears key benefits. These include the potential for increased reliability and precision, and reduced error frequency arising from removal of operator tedium.

In a first order, drug discovery strategies can be divided into the following parts:

i) accessing structural diversity either from natural sources or from synthetic approaches,
ii) sample preparation and handling,
iii) the screening based on new and significant biological assay systems,
iv) the profiling of so-called hits from the primary screening, and
v) data management.

Today, a number of pharmaceutical companies dealing with HTS reach a turnover of more than 15 different assay systems a year, in which 300,000 samples or even more are tested. This confronts the scientist with more than 5,000,000 data points which point to the need for efficient automation at all stages of HTS, even in data collection, data quality control and analysis. Robots, especially large systems integrated with multiple peripheral devices, are prominent at present.

In order to take real advantage of HTS, access to high sample numbers covering a broad range of low-molecular mass diversity is essential. The answer of organic chemistry to high capacities in testing was combinatorial chemistry. The answer for natural product chemistry is still pending. Combinatorial biochemistry aims at enhancing molecular diversity of natural products by altering and combining of biosynthetic genes. Although, these promising strategies may probably contribute substantially to HTS test sample supply in future, improvements that can be realized more rapidly are needed urgently to decrease the delay of innovation, and to focus attention of the management boards in the pharmaceutical companies again towards natural products based test sampling.

2 Natural Product-derived Drugs on the Market

The accidental discovery of penicillin in 1928 by Alexander Fleming and its introduction in 1941/42 as an efficient antibacterial drug bearing no substantial side effects revolutionized medicinal chemistry and pharmaceutical research by stimulating completely new strategies in industrial drug discovery. Penicillin was discovered by chance from the culture broth of *Penicillium notatum* as a bioactive principle inhibiting growth of gram-positive bacteria. In contrast to plants microorganisms were not known to biosynthesize secondary metabolites useful in medicinal application until the discovery of the penicillins. In the following decades microorganisms attracted considerable attention as a completely new source for pharmaceuticals. Microbial metabolites proved available at quantities of up to hundreds of kilograms by fermentation technology. From the screening of a huge number of microbial extracts an unexpected diversity of natural compounds performing a broad variety of biological activities became evident.

Microorganisms turned out to be an unlimited source for both, potential drugs, and agrochemicals [15–21]. Manufacturing can be performed without interfering with bioprospecting issues that today considerably hamper industrial drug development from phytogenic or animal sources. Clinical studies on drug approval require kilogram amounts of the pure active compounds. Chemical synthesis cannot solve the problem in the cases of complex structures, as it was shown by the well known examples of the phytogenic drugs morphine and codeine from *Papaver somniferum*, quinine from *Cinchona* species, atropine from *Atropa belladonna*, reserpine from *Rauwolfia* species, vincristine from *Catharanthus roseus*, or the cardiac glycosides from *Digitalis purpurea*. However, in various cases valuable precursors from plants can be made accessible at moderate prices thus contributing to improved manufacturing processes, supplemented by chemical synthesis, and biocatalysis or biotransformation, respectively.

Precursors are sometimes of considerable advantage because they also give access to artificial analogs or derivatives which may exhibit a superior spectrum of properties referring to the demands for a successful drug development and application. For instance, the manufacture of oral contraceptives and other steroid hormones based on starting materials of plant origin such as diosgenin from *Dioscorea species* and hecogenin from *Agave sisalana*.

Recently, despite all obstacles hampering large scale production of phytogenic drugs, plant metabolites succeeded again in reaching the pharmaceutical market: The urgent demand for improved chemotherapeutics in anticancer treatment forced the approval of Taxol (Bristol-Myers Squibb), its analog Taxotère (Rhône-Poulenc Rorer), topotecan, and irinotecan (Sect. 2.5).

2.1 Antibiotics

Since the introduction of the first cephalosporin antibiotic in 1962, every year several new *β*-lactam antibiotics of the cephalosporin-type have reached the market. Continual improvement in the activity spectrum is necessary to overcome resistance phenomena caused by the widespread application of antibiotics. Referring to the "hit"-list of the commercially most successful therapeutics, three *β*-lactam antibiotics range among the top twenty. These are cefaclor by E. Lilly, ceftriaxone by Roche, and clavulanic acid as a most efficient *β*-lactamase inhibitor marketed by Smith Kline-Beecham in combination with the semi-synthetic penicillin derivative amoxicillin under the trade name of Aug-

Table 1. History of commercialization of modern drugs derived from nature

Year of introduction	Drug	Commercialized as (m = microbial metabolite,) p = plant metabolite	Indication	Company
1826	manufacturing of morphine	natural compound (p)	analgesic	E. Merck
1899	acetylsalicylic acid (Aspirin)	synthetic analog (p)	analgesic, antiphlogistic, etc.	Bayer
1941	penicillin	natural compound (m)	antibacterial	Merck
1964	first cephalosporin antibiotic (cephalothin)	semi-synthetic derivative based on 7-ACA (m)	antibacterial	Eli Lilly
1983	cyclosporin A	natural compound (m)	immunosuppressant	Sandoz
1987	artemisinin	natural compound (p)	antimalaria	Baiyunshan
1987	lovastatin	natural compound (m)	antihyperlipidemic	Merck
1988	simvastatin	semi-synthetic derivative (m)	antihyperlipidemic	Merck
1989	pravastatin	semi-synthetic derivative (m)	antihyperlipidemic	Sankyo/ BMS
1990	acarbose	natural compound (m)	antidiabetic (type II)	Bayer
1993	paclitaxel (Taxol	natural compound (p) as a semi-synthetic derivative	anticancer	BMS
1993	FK 506 (tacrolimus)	natural compound (m)	immunosuppressant	Fujisawa
1994	fluvastatin	synthetic analog (m)	antihyperlipidemic	Sandoz
1995	docetaxel (Taxotère)	semi-synthetic derivative (p)	anticancer	Rhône-PR
1996	topotecan, irinotecan	semi-synthetic derivatives[a] (p)	anticancer	SKB, Pharmacia & Upjohn
1996	miglitol	synthetic analog (m, p)	antidiabetic (type II)	Bayer

[a] Irinotecan was launched first in Japan in 1994 by Yakult Honsha and Daiichi Pharmaceutical. BMS = Bristol-Myers Squibb, Rhône-PR = Rhône-Poulenc Rorer, SKB = SmithKline Beecham.

mentin [20, 22]. Regarding the hit-list of antibacterials, the only competitive synthetics to the β-lactams, so far, are the quinolone carboxylic acids affecting the bacterial topoisomerase (DNA gyrase) [23]. The top-selling drug ciprofloxacin was introduced under the trade name of Ciprobay by Bayer in 1986.

Further antibiotics, mainly derived from actinomycetes, are used for special applications in human and veterinary medicine [20]. These compounds have numerous chemical structures. The macrolides, tetracyclines, aminoglycosides, glycopeptides, and ansamycins for instance are used in antibacterial treatment whereas the anthracyclines reached the market to supplement anticancer chemotherapy. The fairly toxic polyether-type antibiotics are preferably used as anticoccidial agents. Due to the dramatically increasing resistance of clinical important bacterial strains new targets for the discovery of novel types of antibacterial agents are urgently needed.

2.2 Immunosuppressant Drugs

The cyclic peptide cyclosporin A was identified at Sandoz in Switzerland as an antifungal agent isolated from culture broths of the fungus *Tolypocladium inflatum*, former classified as *Trichoderma polysporum* [24]. In the course of more detailed pharmacological studies applying improved in vivo animal models cyclosporin A performed remarkable immunosuppressive properties [25, 26]. As a result of its superior efficiency, cyclosporin A was commercialized under the trade name Sandimmune only seven years after it had been first reported. For the first time, transplantation surgery was provided with a drug that prevented organ rejection effectively without substantially affecting the immune response giving protection against bacterial infections. Considering that patients of successful transplantation surgeries have to take cyclosporin A for the rest of their life this issue is of great importance. Cyclosporin A is still in the top twentyfive drugs worldwide, although it was introduced as early as 1983. However, in spite of great efforts made by Sandoz and numerous competitors to develop an improved immunosuppressant, cyclosporin A, is still in the number one position in immunosuppressive drug sales [27].

Studies on the mode of action of cyclosporin A revealed interaction with intracellular signal transduction pathways leading to the inhibition of interleukin-2 production in T-cell cultures. Subsequent screening for inhibitors of interleukin-2 release at Fujisawa in Japan resulted in the discovery of the macrocyclic lactame FK 506 from the culture broth of *Streptomyces tsukubaensis* in 1984 [28, 29]. Structural similarities to rapamycin, previously described as an antifungal metabolite from *Streptomyces hygroscopicus* [30], gave rise for the reevaluation of the biological activities of rapamycin. Detailed studies on the mechanism of action of rapamycin, FK 506, and cyclosporin A yielded completely new insights into eukaryotic cell signaling, and led to the discovery of a new and unusual class of proteins called immunophilins. Although FK 506 and rapamycin bind to the same protein, called FK binding protein (FKBP), the complexes formed trigger different signaling pathways. In T-cell cultures the FK 506/FKBP complex binds to calcineurin thus preventing its interaction with calcium ions that regularly bind to calcineurin after activation of the T-cell receptor. Subsequent to the inhibition of calcineurin activation, the signal transduction leading to the transcription of gene encoding, e.g. for interleukin-2, is prevented. As a consequence, T-cell proliferation does not occur. However, the rapamycin/FKBP complex does not bind to calcineurin, but inactivates protein kinases that respond to interleukin-2 receptor activation by stimulating T-cell proliferation. Cyclosporin A was shown to bind to another type of immunophilin called cyclophilin. Surprisingly, the cyclosporin A/cyclophilin complex binds to calcineurin at the same site as FK 506/FKBP does, thus eventually triggering the same steps of the cell signaling cascade [31].

Although, at the time of discovery FK 506 was reported as a new immunosuppressant agent being 100-fold more potent than cyclosporin A, efforts to displace cyclosporin A by FK 506, commercialized as tacrolimus, failed until today. It is worth mentioning that FK 506 stimulated the setup of the venture capital funded US company Vertex that aims at a designed drug to treat HIV infections starting from FK 506 as a lead compound.

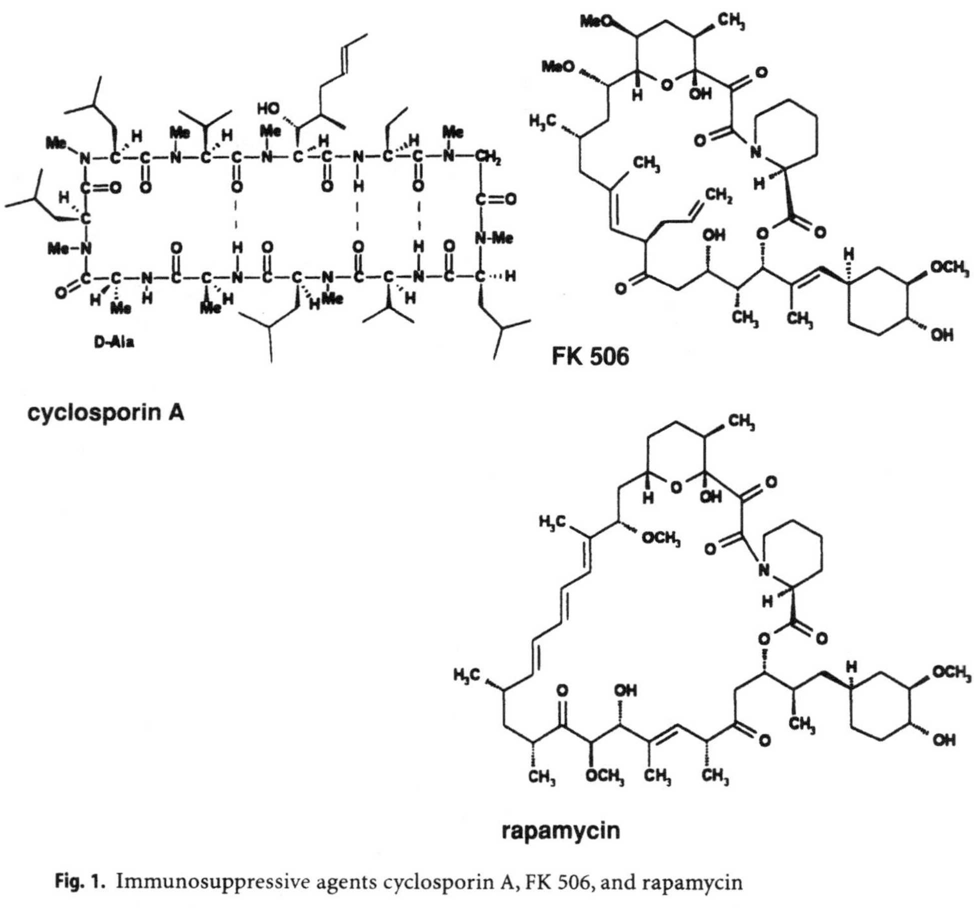

Fig. 1. Immunosuppressive agents cyclosporin A, FK 506, and rapamycin

2.3 Antihypercholesterolemic Drugs

For several years, lovastatin represented the commerically most successful drug from nature. Named mevinolin, it was discovered as a metabolite from *Aspergillus terreus* cultures at the US company Merck by a target-directed screening for inhibitors of 3-hydroxy-3-methylglutaryl coenzyme A (HMG-CoA) reductase, a key enzyme of cholesterol biosynthesis [32]. Cholesterol, which in humans more than one-half of the total body cholesterol is derived from its *de novo* biosynthesis in the liver, is the major component of atherosclerotic plaques built up as fatty deposits on the inner walls of arteries, thus contributing to arteriosclerosis and coronary heart diseases [33].

A number of mevinolin like HMG-CoA reductase inhibitors comprising the new type of a δ-lactone moiety linked to a decalin sidechain were isolated from fungal strains such as *Penicillium, Aspergillus,* and *Monascus spec.* by various groups and companies during the late 1970s and early 1980s [34–40]. Mevinolin and its analogs do not affect any other enzyme in the cholesterol biosynthesis except HMG-CoA reductase (K_i value of 6.4×10^{-10}; rat liver enzyme [32]) thus realizing a completely new therapeutic principle to treat hypercholesterolemia. Mevinolin was the first HMG-CoA reductase inhibitor to be commercialized. It was approved under the generic name of lovastatin (Mevacor) by Merck in 1987. Later on, numerous lovastatin analogs developed by Merck and their competitors followed. In 1994, simvastatin (Zocor), introduced by Merck in 1988, took over the top position in the natural product based drugs sales list. Sankyo's mevastatin, isolated and described prior to mevinolin, had to be modified to reach the market under the generic name of pravastatin (Mevalotin) co-marketed by Bristol-Myers Squibb in 1989. Pravastatin represents the first HMG-CoA inhibiting drug with a opened ring dihydroxycarboxylic acid structure.

In contrast to pravastatin, lovastatin and its δ-lactone analogs have to be metabolically activated to bind to the target site of the HMG-CoA reductase. Cleavage of the δ-lactone moiety in the liver and in the cell culture system, respectively, leads to the 3,5-dihydroxyvaleric acid form that mimics the structure of the proposed endogenous substrate consisting of an intermediate formed by mevalonic acid linked to coenzyme A. Thus, the dihydroxycarboxylic acid form of lovastatin, and its analogs act as competitive inhibitors comprising about 10,000- to 20,000-fold increased affinity to the target enzyme. Numerous reports on the mode of action and structure/activity relationships have been published [41–48]. The replacement of the decalin structure by certain substituted biaryl analogs retains potency, thus leading e.g. to the synthetic drugs fluvastatin marketed by Sandoz [49], and Bay W 6228 developed at Bayer [50]. Presumably, manufacturing costs of both compounds are significantly reduced relative to the fermentation based drugs lovastatin, simvastatin and pravastatin.

HMG-CoA reductase inhibiting drugs act within a highly complex bio-control system, and finally result in a plasma cholesterol decrease of up to 40%. Sales rates of lovastatin and its analogs are still growing, and in 1994 their sales reached more than US $ 3.6 billion. Although lovastatin and its analogs exhibit only rare side-effects, a lot of people criticize the use of antihypercholesterole-

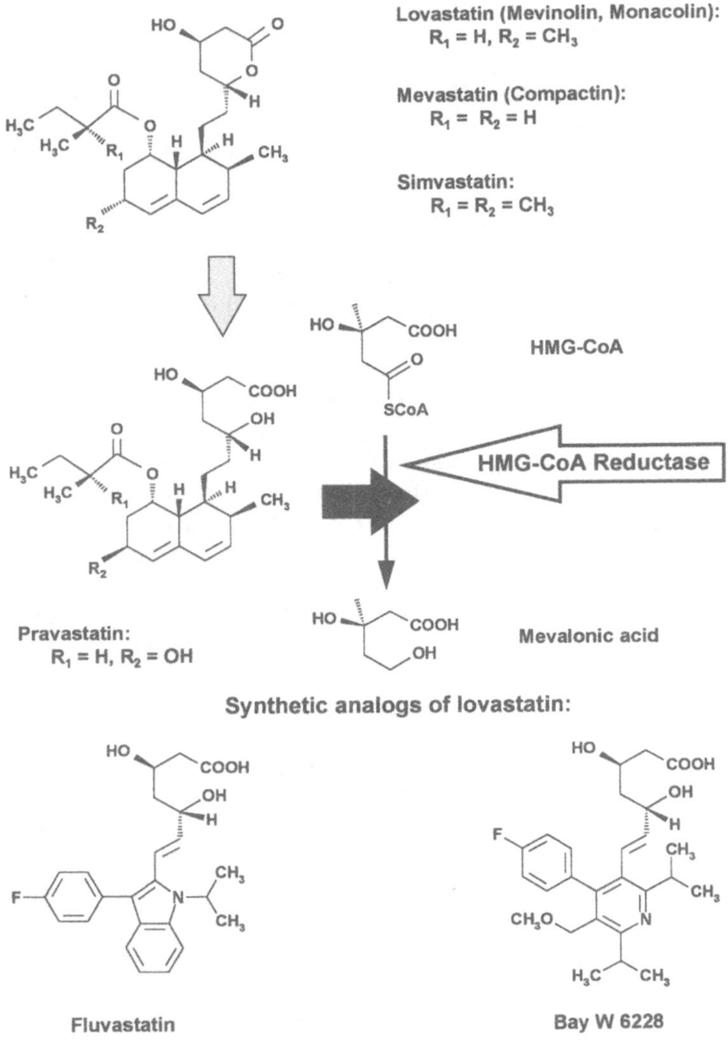

Fig. 2. Inhibition of cholesterol biosynthesis by lovastatin and its analogs

mic drugs instead of encouraging better nutrition. However, HMG-CoA reductase inhibitors seem to be extremely helpful therapeutics for patients suffering from hereditary hyperlipidemia causing severe coronary heart diseases, and cardiac infarction of even very young people.

2.4 Antidiabetic Drugs

Acarbose was discovered in a target-directed screening from the culture broth of a neglected genera of actinomycetes called *Actinoplanes spec.* by researchers at the German company Bayer in the mid 1970s [51]. Their screening aimed at inhibitors of oligo- and polysaccharide degrading enzymes towards an orally available drug contributing to Diabetes mellitus type II (non-insulin-dependent) therapy [52, 53]. Enzyme inhibitors of this type were expected to reduce food intake-caused blood glucose elevation by retarding degradation of nutrient oligo- and polysaccharides in the intestine. And in fact, acarbose can be applied effectively when a diet alone fails to adequately control blood glucose levels. Furthermore, it was shown that insulin dosing of patients suffering from Diabetes mellitus type I (insulin-dependent) can also take advantage of acarbose because the exaggerated rise in blood glucose level after meals can be reduced.

Acarbose was first launched by Bayer in Germany under the trade name of Glucobay in 1990. The pseudo-disaccharide moiety of the pseudo-tetrasaccharide acarbose strongly inhibits the intestine located enzymes of the α-glucosidase type by mimicking a disaccharide part of the starch molecule or related nutrient polymers which functions as the natural substrate of the target enzymes. The glucose moiety in acarbose is replaced by an unsaturated cyclohexitol (C_7N-unit), thus preventing degradation of the pseudo-tetrasaccharide as a competitive enzyme inhibitor [54].

Bayer succeeded in reducing manufacturing costs of acarbose dramatically, presumably mainly by increasing yields and shifting the fermentation process to the desired homologous compound acarbose. Efforts to replace the fermentation by chemical synthesis of a structurally simplified analog of the tetrasaccharide failed. However, the profile of oral diabetes treatment by Bayer is actually complemented by the monosaccharide miglitol, an additional α-glucosidase inhibitor [55, 56] which was synthesized starting from the naturally occurring 1-deoxynojirimycin as a lead structure. Interestingly, 1-deoxynojirimycin can be isolated from both, culture broths of streptomycetes and the bark of mulberry trees.

2.5 Anticancer Drugs

The example of the anticancer drug Taxol, introduced by Bristol-Myers Squibb in 1993, demonstrates how to overcome the obstacles affecting the development

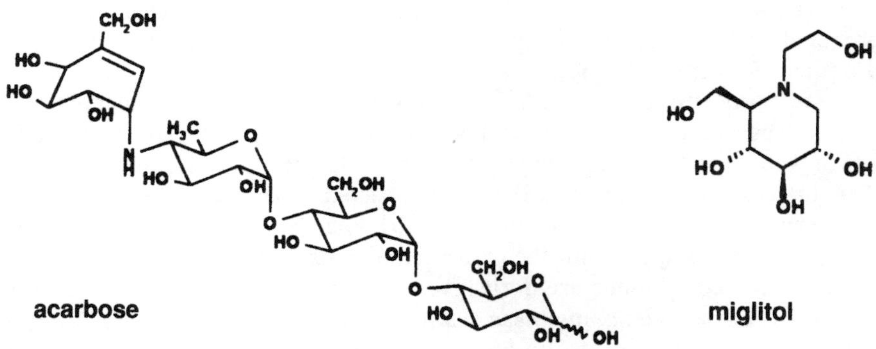

Fig. 3. Antidiabetic drugs acarbose, and miglitol

of a plant-derived drug nowadays. Taxol was discovered at the US National Cancer Institute (NCI) in the late 1960s in the course of an in vitro antitumor drug discovery program using human tumor cell lines. Screening of more than 110,000 samples derived from more than 35,000 plant genera collected worldwide resulted in the isolation and structure elucidation of Taxol from the bark of the Californian yew tree *Taxus brevifolia* [57].

Taxol performed outstanding activities against various tumor cell lines, but clinical studies did not take place due to insufficient amounts of plant material available. Only after identification of the unique mode of action of Taxol [58, 59], both the NCI and pharmaceutical industry were attracted to pursue profiling Taxol as a completely new anticancer drug [60]. Taxol was shown to affect mitosis by increasing microtubuli assembling and decreasing microtubuli degradation rates. In order to enable clinical testings, for years, the only way to get access to Taxol was to shed the bark of more than sixty year old Californian yew trees which by the harvesting procedure were damaged irreversibly. However, worldwide objections by environmentalists were rejected by the US authorities.

In the meantime, the problem of Taxol supply seems to be solved. The European yew tree *Taxus baccata* was identified as a renewable source of Taxol-type metabolites. By harvesting and extracting the needles baccatin III or 10-deacetyl baccatin III are obtained in good yields without injuring the trees substantially and are used as precusors of Taxol synthesis. Starting from these compounds, not only Taxol but also analogs such as Taxotère (docetaxel) were synthesized in satisfying yields. Taxotère was recently launched by the French company Rhône-Poulenc Rorer. The company was free to apply for a product patent to improve commercial issues. However, Bristol-Myers Squibb only achieved trade name protection for Taxol at the time of marketing, because the chemical structure of Taxol had already been published in 1971. Thus, the generic name of Taxol was designated paclitaxel. Actually, Taxol is approved for treatment of breast and ovarian cancer.

Efforts to get access to improved manufacturing processes or improved application properties of paclitaxel resulted in two different pathways for the total synthesis of the natural product, as well as analogs. Furthermore, paclitaxel and/or precursors can be obtained under optimized conditions from cell cultures of *Taxus media* generated by hybridizing *Taxus baccata* and *Taxus cuspidata* in overall yields of about 130 mg l^{-1} within two weeks [61]. Paclitaxel can also be obtained by the culture of appropriate microbial strains isolated from paclitaxel producing yew trees [62–64].

A few years after the introduction of Taxol in 1996, further phytogenic anticancer drugs were launched to treat advanced cancers. Topotecan, marketed by Smith Kline Beecham under the trade name of Hycamtin, was approved by the FDA to treat ovarian cancers that have resisted other chemotherapy drugs. Furthermore, irinotecan was introduced by Pharmacia & Upjohn under the trade name of Camptosar for the treatment of metastatic cancer of the colon or rectum. Both compounds are derivatives of camptothecin which was isolated from the Chinese tree *Camptotheca acuminata*, well known in Chinese Traditional Medicine for anticancer treatment [65]. Isolation of the bioactive principle camptothecin and its structure elucidation had already been performed in 1966

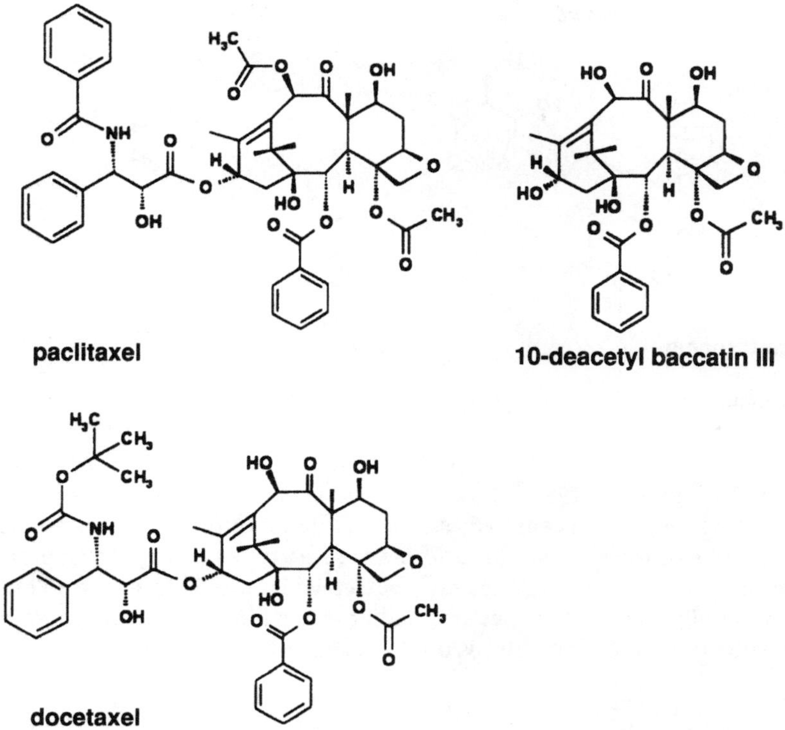

Fig. 4. Anticancer drugs paclitaxel and docetaxel. Both compounds can be obtained by chemical structure modification of baccatin III or 10-deacetyl baccatin III

[66]. However, only recognition of the topoisomerase I as the site of action stimulated the pharmaceutical industry to develop camptothecin as a novel anticancer drug. In contrast to the well known anthracyclin-type antitumor therapeutics which bind to topoisomerase II, camptothecin was the first compound interacting with topoisomerase I. It binds to the complex formed by topoisomerase I and DNA which starts DNA replication, thus stabilizing the enzyme/DNA complex and preventing cell proliferation [67]. Chemical modification of camptothecin, however, did not result in substantially optimizing efficiency but in improving solubility in water by replacing the hydrogen atoms at position 9 and 10 leading to topotecan, or at position 7 and 10 leading to irinotecan, respectively [68, 69].

2.6 Drugs for various Applications

Traditional Chinese Medicine gave rise to new plant-derived anticancer drugs as well as to therapeutics for the treatment of drug-resistant types of malaria tropica [70]. Artemisinin (Qinghaosu), the bioactive principle of *Artemisia annua*, was identified in 1972. Its unique structure was proven by X-ray analysis

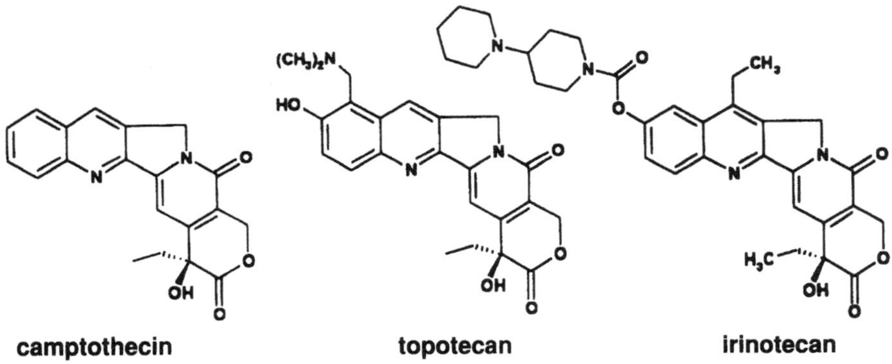

Fig. 5. Camptothecin-derived drugs topotecan, and irinotecan

and total synthesis in 1983 [71]. Artemisinin is actually applied in Asian countries in cases of malaria caused by strains of *Plasmodium falsiparum* resistant to therapeutics of the chloroquine and mefloquine type [72, 73]. Artemisinin exhibits only moderate acute toxicity but considerable side effects. Thus, for approval in European countries artemether which is a semi-synthetic derivative of artemisinin indicating improved tolerance has been developed [74].

3 Clinical Candidates Derived from Natural Sources

3.1 Clinical Candidates from Microorganisms

Due to failures in cancer therapy, numerous compounds derived from various natural sources are under preclinical or clinical trials in order to study their therapeutic value. One example is the extremely toxic enediynes isolated from actinomycetes strains. Featuring a *cis*-configurated double bond adjacent to two triple bonds within a macrocyclic ring system as a part of a highly complex molecule, the enediyne structure was first described for neocarzinostatin in 1965 [75]. Other members of the enediyne antibiotic class, the esperamicins [76], the calicheamicins [77], both featuring a unique allylic trisulfide attached to a bridgehead carbon atom, and dynemicin A [78] were discovered in the mid- to late 1980s. All these compounds appear to share two modes of action which comprise DNA intercalation to the minor groove, and second a thiol or NADPH triggered reaction generating radicals that cleave DNA [79]. A Bergman reaction represents the key step that leads to the formation of a highly reactive diradical being responsible for DNA strand scission by abstracting hydrogen atoms at the C-1′, C-4′ or C-5′ position of the DNA sugar-phosphate residues within the minor groove. In vitro studies implicate sequence selectivity for DNA strand scission ranging from base preferences of the esperamicins at thymidine to cytidine for the calicheamicins, and guanine for dynemicin A, respectively.

However, as there is no free DNA in the cell, and the real target for DNA-interactive agents is the chromatin, it is not clear whether in vitro studies can contribute to understanding of cellular DNA sequence selectivity. Various synthetic approaches and mechanistic studies assisted by computer aided molecular design (CAMD) have been described aiming at an optimization of the pharmacological properties of the enediyne antibiotics [80]. It remains to be seen whether enediynes, either natural or designed, will become a useful supplement to the arsenal for the treatment of cancer.

A further example emphasizes the relevance of microbial metabolites in the development of novel antitumor drugs. CC 1065 was discovered at the Upjohn Company as an extremely cytotoxic metabolite from the culture broths of *Streptomyces spec.* [81, 82]. Hepatotoxic properties with LD_{50} of 9 mg/kg (mice i.v.) prevent therapeutic use of the natural product. Investigation on the molecular mechanism of action together with chemical modification of the basic structure [83–86] resulted in the bifunctional synthetic analog bizelesin [87] that bears an urea moiety linked symmetrically to two DNA binding and alkylating groups. Bizelesin is unique among the analogs with bifunctional alkylating capability due to two chloromethyl moieties which are converted to the cyclopropyl alkylating species that interacting DNA by producing interstrand cross-links [88–91]. As bizelesin did not lead to delayed deaths when applied in therapeutic doses to mice, and due to its breadth of antitumor activity, potency, unique mechanism of action, and lack of cross-resistance with other alkylating agents,

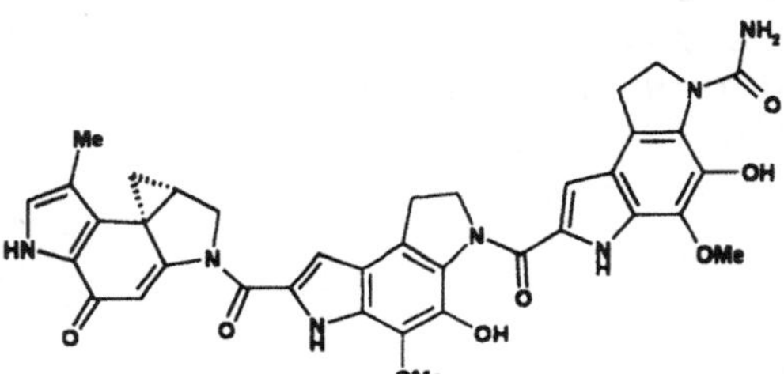
CC 1065

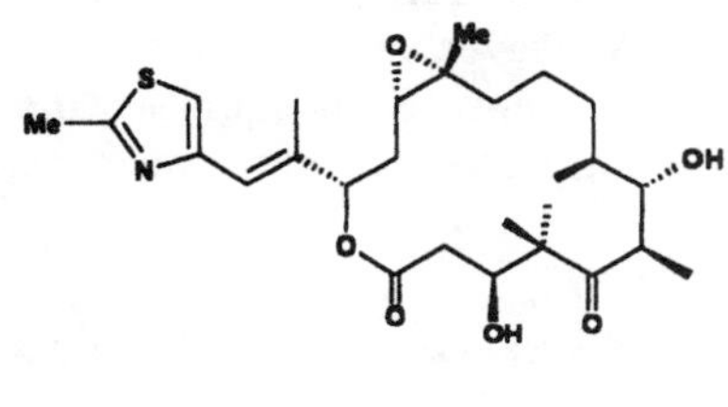
epothilone B

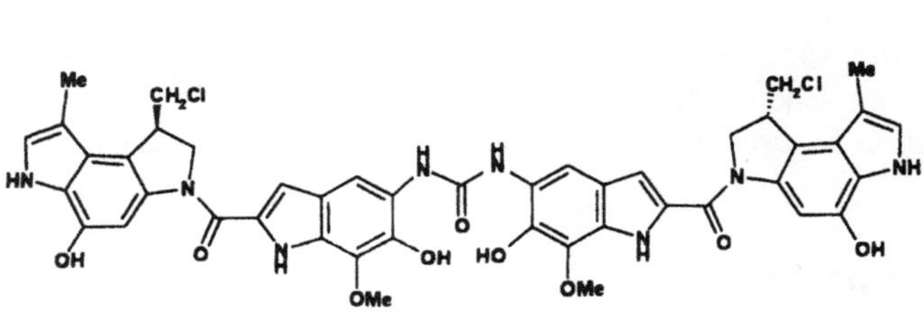
bizelesin

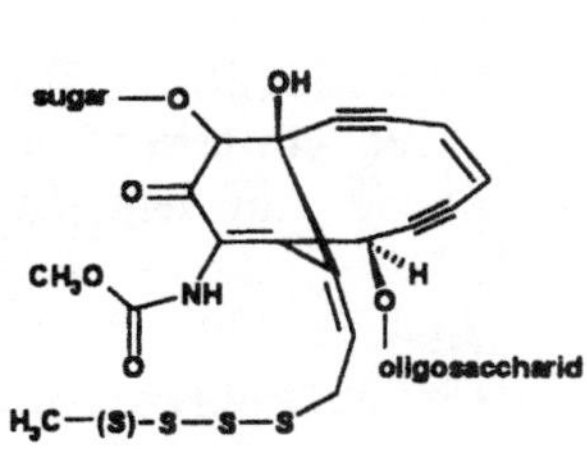

esperamicins

Fig. 6. Microbial metabolites esperamicin, CC 1065, and epothilone comprising potential anticancer activity. Starting from CC 1065 as a lead structure bizelesin was synthesized.

bizelesin was selected by the National Cancer Institute and the Upjohn Company for development in clinical trials in the USA [92].

Epothilone was first discovered by Höfle and Reichenbach at the GBF in Germany in 1987 within a screening for antifungal agents from the cultures of the myxobacterium *Sorangium cellulosum* [94]. In 1995, Merck & Co. reported on its interaction with the microtubuli of cells in a paclitaxel like manner [93]. Although epothilone does not reveal any structure similarity to paclitaxel [95] it turned out to be superior to paclitaxel referring to potency in cell culture assays of multi-drug-resistant tumor cell lines. The microbial metabolite epothilone is accessible in almost unlimited quantities by fermentation. Furthermore, it overcomes solubility problems in water that substantially reduce paclitaxel compliance. Total synthesis towards epothilone and analogs as well as modification reactions [96–98] starting from epothilone aim at epothilone derivatives with improved performance. In 1997, Bristol-Myers Squibb has signed a contract with the GBF for extended investigation towards an epothilone based novel anticancer drug supplementing the therapeutic range of Bristol-Myers Squibb's pharmaceutical Taxol.

A natural products screening for inhibitors of the squalene synthase, a late enzymatic step in the biosynthetic pathway leading to cholesterol, gave rise to the parallel discovery of novel highly oxygenated types of fungal metabolites called zaragozic acids (Merck [99]), and squalestatins (Glaxo [100], and Mitsubishi Kasei [101]). The compounds exhibit IC_{50} values in the nanomolar range (rat enzyme). Although cholesterol biosynthesis is substantially reduced in various animal models by oral application of zaragozic acids/squalestatins their toxicity profile comprising accumulation of farnesyl-derived dicarboxylic acids presumably will prevent medicinal use as antihypercholesterolemic drugs. However, due to their structural similarity with farnesyl pyrophosphate the zaragozic acids/squalestatins also interact with farnesyl transferase, a promising target for new anticancer therapeutics [102].

3.2
Clinical Candidates from Plants

The impact of Traditional Chinese Medicine on modern western drugs is demonstrated not only by commercialized therapeutic agents such as camptothecin or artemisinin but also by clinical candidates like huperzine A, an acetylcholinesterase inhibitor to treat Alzheimer's disease [103]. Huperzine A can be isolated from both, *Huperzia serrata*, and *H. selago* [104, 105]. Because huperzine A is produced only at very low levels a synthetic approach has been developed in order to provide sufficient quantities for preclinical toxicology studies and clinical trials.

3.3
Clinical Candidates from Marine Environment

Recent trends in drug discovery from natural sources emphasize investigation of the marine environment yielding numerous, often highly complex chemical structures [106–108]. So far, in most cases in vitro cultivation techniques towards the supply with sufficient quantities for exhaustive biological activity profiling and clinical testing are missing. Focus on marine biotechnology is actually forced by results indicating that marine microorganisms are substantially involved in the biosynthesis of marine natural products isolated from collected macroorganisms such as invertebrates.

It remains open whether marine natural products will play an important role in drug discovery and development in future. Today, toxic principles dominate the spectrum of biologically active metabolites isolated from marine sources. This fact may partly be due to the major application of cytotoxicity-directed screening assays. However, it has to be considered that defence strategies are necessary to survive in the highly competitive marine environment thus resulting in a tremendous diversity of highly toxic compounds affecting numerous targets involved in eukaryotic cell signaling processes.

The extreme toxic potential of marine metabolites often prevents their application in medicine. However, a number of metabolites proved to be valuable tools in biochemistry, cell and molecular biology. For instance the neurotoxic maitotoxin [109–112] (interaction with extracellular calcium; enhancement of calcium influx [113]), the neurotoxic brevetoxin B [114] (interaction with the binding-site-5 of voltage-sensitive sodium channels [115]), tetrodotoxin and saxitoxin (voltage clamp analysis to study sodium channels and excitatory phenomena [116]; tetrodotoxin abolishes brevetoxin B activity [117]), okadaic acid [118–120] (analysis of phosphorylation and dephosphorylation processes in eukaryotic cell metabolism [121]), and palytoxin (stimulation of arachidonic acid metabolism synergistically with TPA-type promoters [122]).

Various other marine natural products exhibit considerable toxic potency which actually gives hope to clinical use mainly in anticancer therapy:

i) Didemnin B, isolated from the ascidian *Trididemnum solidum*, was the first marine metabolite which reached human clinical trials. To provide sufficient quantities total synthesis of the depsipeptide had to be realized [123]. The didemnins exhibit anticancer as well as antiviral and immunosuppressant activities by inhibiting protein biosynthesis without interacting with the DNA, and independent of the cell cycle [124, 125].
ii) The bryostatins isolated from the bryozoan *Bugula neritina* [126] affect protein kinase C at picomolar concentrations [127, 128]. At present, the potential use of bryostatin 1 in anticancer treatment is studied in clinical trials. Multi-gram quantities have been made accessible following current good manufacturing practices (cGMP) [129].
iii) The alkaloid type ecteinascidin 743 which first has been isolated by Rinehart and coworkers from the mangrove associated ascidian *Ecteinascidia turbinata* [130] expresses potent broad spectrum cytotoxic properties by reacting with a guanine rich site in the DNA's minor groove [131]. Clinical studies sponsored by the company PharmaMar are currently underway.

However, clinical studies on marine metabolites are often hampered by insufficient supply of material, as it was reported for dolastatin 10 from the sea hare *Dolabella auricularia* [107]. In the case of manoalide, obtained from the sponge *Luffariella variabilis*, it remains open whether its use will be restricted to an application in studying inflammatory processes on the cell level or whether it

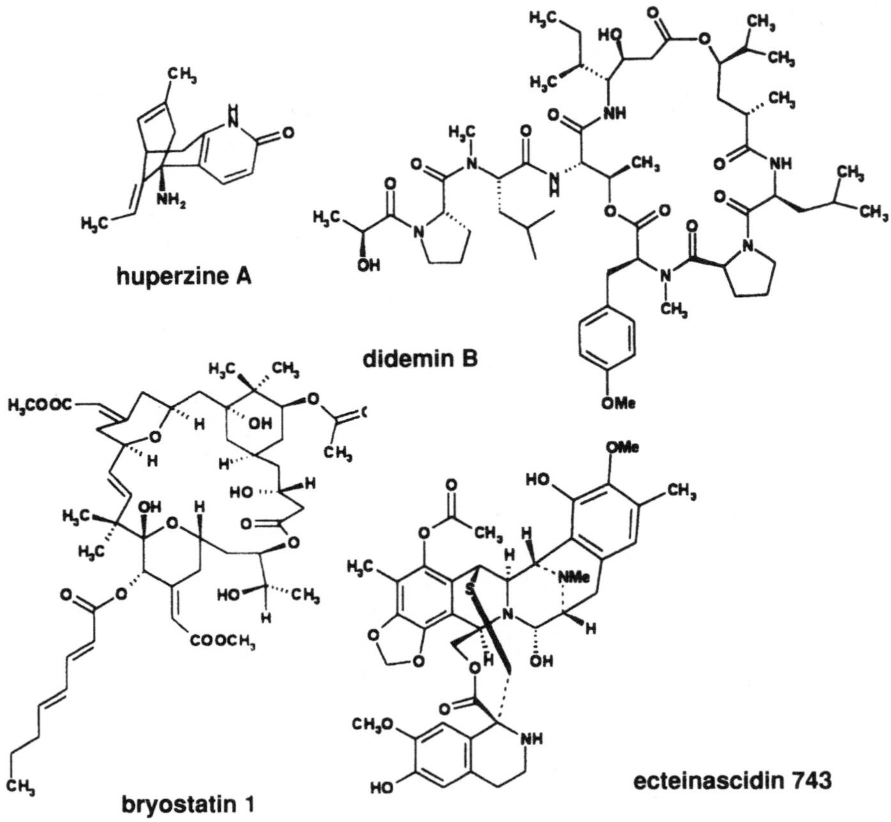

Fig. 7. Clinical candidates from plants and marine environment: Huperzine A, didemnin B, bryostatin 1, and ecteinascidin 743.

will give rise to synthetic analogs to treat inflammation by interacting with phospholipase A_2 [132, 133].

Recently, Pseudopterosin C, a diterpene riboside isolated from the gorgonian corals *Pseudopterogorgia bipinata* and *P. elisabethae* [134] was launched by Estée Lauder as an additive to anti-aging cosmetics. It exhibits potent anti-inflammatory and analgesic activities by reversibly affecting both, lipoxygenase and phospholipase A_2 [135].

4 Exploitation of Structural Diversity

4.1 Approaches to Access Structural Diversity from Nature

Today, more than 30,000 diseases are clinically described. Less than one third of these can be treated symptomatically, and only a few can be cured [136]. Consequently, there exists a strong interest in getting access to new therapeutic agents which is one driving force for advanced drug discovery strategies. New chemical entities are requested in order to enable therapeutic innovation. An analogs situation is observable towards new candidates for agricultural application. Both, combinatorial chemistry and the exploitation of structural diversity derived from natural sources contribute to improved lead discovery. In particular, low-molecular mass natural products from bacteria, fungi, plants, and invertebrates either from terrestric, or marine environments perform unique structural diversity. In order to get access to this outstanding molecular diversity various strategies like the (target-directed) biological, physico-chemical, or chemical screening have been developed. Due to the commercial superiority of microbial metabolites this section will focus on chemical diversity derived from microorganisms.

In the biological screening the selection criterion usually is a wanted biological effect aiming at a defined pharmaceutical application (target-directed biological screening) [137]. Biological screening has been developed to a powerful concept which integrates and makes use of recent findings in molecular and cell biology mounting in high-throughput screening attempts (HTS). Today, success in drug discovery and development obviously depends on the therapeutic value of the bioassays running in the primary biological screening, and on the period required for the identification of first promising lead compounds in order to start with lead optimization procedures. Therefore, biological screening strategies are considered to be the most profitable ones and are in widespread practice in pharmaceutical companies.

In contrast to biological screening, the physico-chemical-, and chemical screening approaches *a priori* possess no correlation to a defined biological effect. Here, a selection of promising secondary metabolites out of natural sources is based on physico-chemical properties, or on the chemical reactivity, respectively. In both strategies, the first step is chromatography in order to separate the compounds from the complex mixtures obtained from plants, microorganisms, fungi, or animals. In a second (analytical) step, physico-chemical properties or chemical reactivities of the separated secondary metabolites are analyzed. Both strategies have shown to be efficient supplemental and alternative methods, especially with the aim to discover predominantly new secondary metabolites.

Chemical screening is based on the analysis of the chemical reactivity of secondary metabolites using thin-layer chromatography (TLC) [138–146]. Concentrates from natural sources are applied on TLC-plates, chromatographed with different solvent systems (separation step), and their chemical reactivity is analyzed by making use of defined chemical reagents sprayed directly onto the TLC plates. Selection criteria are both, the chromatographic parameters, and the staining (colorization) behavior after spraying and heating. In a physico-chemical screening selection criteria are the chromatographic behavior (e.g. on C_{18}-colums by using HPLC for the chromatographic step), as well as data from e.g. UV/VIS-, MS-, IR-, or in rare cases NMR spectra. In most cases, coupling techniques of HPLC with the desired analytical method are used.

4.1.1
Physico-Chemical Screening

Mycelium extracts, culture filtrates, or crude extracts of microbial broths as well as samples obtained from plant and animal extraction can be subjected to standardized reversed-phase HPLC by making use of various coupling techniques. Most common is HPLC coupled to a multi-wavelength UV/VIS-monitor (diode array detection, DAD) [147]. Comparison of the data (retention time and UV/VIS-spectra) to reference substances acts as selection criteria. However, success of this strategy depends upon the amount and quality of pure references in the database. Based on the correlation to UV/VIS monitoring the HPLC-DAD screening is well suitable for screening towards metabolites which bear significant chromophores. In combination with the efficient separation via HPLC this screening procedure can advantageously be applied to plant material which exhibits numerous colored compounds. The HPLC-DAD screening has already been successfully applied to natural products screening and led to the discovery of several new metabolites like the naphthoquinone juglomycin Z [148], naphthgeranine F [149], the antibiotically active fatty acid (*E*)-4-oxonon-2-enoic acid [150], the insecticidal acting NK374200 [151], or the quinoxaline group metabolite echinoserine [152].

The data obtained from HPLC-DAD analysis are often helpful in dereplication, e.g. during high-throughput biological screening programs. However, the dependence on a UV/VIS detectable chromophor in the metabolite to be analyzed limits its possible application in principal. Therefore, alternative detection methods like mass spectrometry (LC-MS) [153, 154], or nuclear magnetic resonance (LC-NMR) [155] have been applied. In a further concept, analytical data from HPLC-DAD can advantageously be supplemented and combined with data from HPLC-ESI-MS. An example for a successful screening is the discovery of the polyol macrolides of the kanchanamycin group [156, 157].

4.1.2
Chemical Screening

In order to apply chemical screening on TLC starting from microbial cultures in a reproducable way, standardized procedures for sample preparation and at least 50-fold concentration are required. The obtained concentrates from both, the mycelium and the culture filtrate are analyzed by applying a defined amount to high-performance thin-layer chromatography (HPTLC) silica-gel plates which then are chromatographed using different solvent systems. In a next step the metabolite pattern of each strain is analyzed by making use of visual detection (colored substances), UV-extinction/fluorescence, and colorization reactions obtained by staining with different reagents (e.g. anisaldehyde/sulfuric acid-, naphthoresorcin/sulfuric acid-, orcinol-, blue terazolium-, and Ehrlich's reagent). The advantage of these reagents lies in the broad structural spectrum of metabolites stained.

This procedure mainly focuses on the chemical behavior and reactivity of the components and gives a good visualization of the secondary metabolite pattern produced by each strain. On the basis of reference substances a number of spots on the chromatograms can be classified to be:

i) constituent of the nutrient broth,
ii) frequently formed and thus widely distributed microbial metabolite,
iii) a strain specific compound which merits further attention.

As a consequence, chemical screening results in the visualization of a nearly complete picture of a secondary metabolite pattern which can be characterized to be a metabolic finger-print of an individual strain [158, 159].

Applied to a number of culture extracts from microbes (e.g. streptomycetes) these screening concepts resulted in various structurally new secondary metabolites. In a following step the obtained pure compounds have to be assayed in different biological test systems. The availability of both, sufficient quantities of pure metabolites, and a broad range of biological tests are the critical points for success. As presumed [142], a certain percentage of these secondary metabolites out of chemical screening progams showed striking biological effects and gave reason for further more detailed biological studies towards new lead structures. Chemical screening resulted in compounds out of nearly every structural class known so far [158–192]. Some selected examples are summerized in Fig. 8.

Recently, in our laboratories the sample preparation procedure (adsorption of the metabolites present in the culture filtrate) has been further developed towards automation and additional separation steps using modified Zymark RapidTrace modules. On the basis of new adsorber resins high quality sample preparation is now possible allowing a more detailed TLC-analysis of the metabolites produced. This leads to an improvement in the separation of overlapping spots and therefore to a higher reliability of the database assisted analysis. On the other hand, sample application onto the TLC-plates is possible via commercially available automated spotting stations. In consequence, chemical screening was adopted to the 96-well format, thus efficiently supplementing biological screening attempts.

Furthermore, the analytical power (fast, cheap, parallel, easy-to-handle, use of UV/VIS and staining reagents) allows one to add physico-chemical information on "hit-lists" out of target-directed screening approaches with samples from natural sources. Therefore, integration of TLC-analysis in the course of secondary biological screening and hit-verification is expected to be a remarkable tool in lead-structure finding strategies, e.g. for the localization of the active principle, for fast and efficient dereplication, and for speed-up isolation and purification procedures. More recently, coupling techniques of mass spectrometry, or even TLC-FID coupling [193] have been described which obviously can be integrated in chemical screening concepts.

In comparison to a TLC-based screening both the quality and sensivity of HPLC separations of a physico-chemical screening via HPLC is better. On the other hand, TLC allows a parallel, quick, and cheap handling of samples, and is superior in the mode of detection (UV/VIS and staining). As well as eluted compounds out of HPLC separations, spots from TLC can easily be subjected to subsequent physico-chemical analysis (MS, IR, NMR etc.) via scrape off and elution from the silica gel materials.

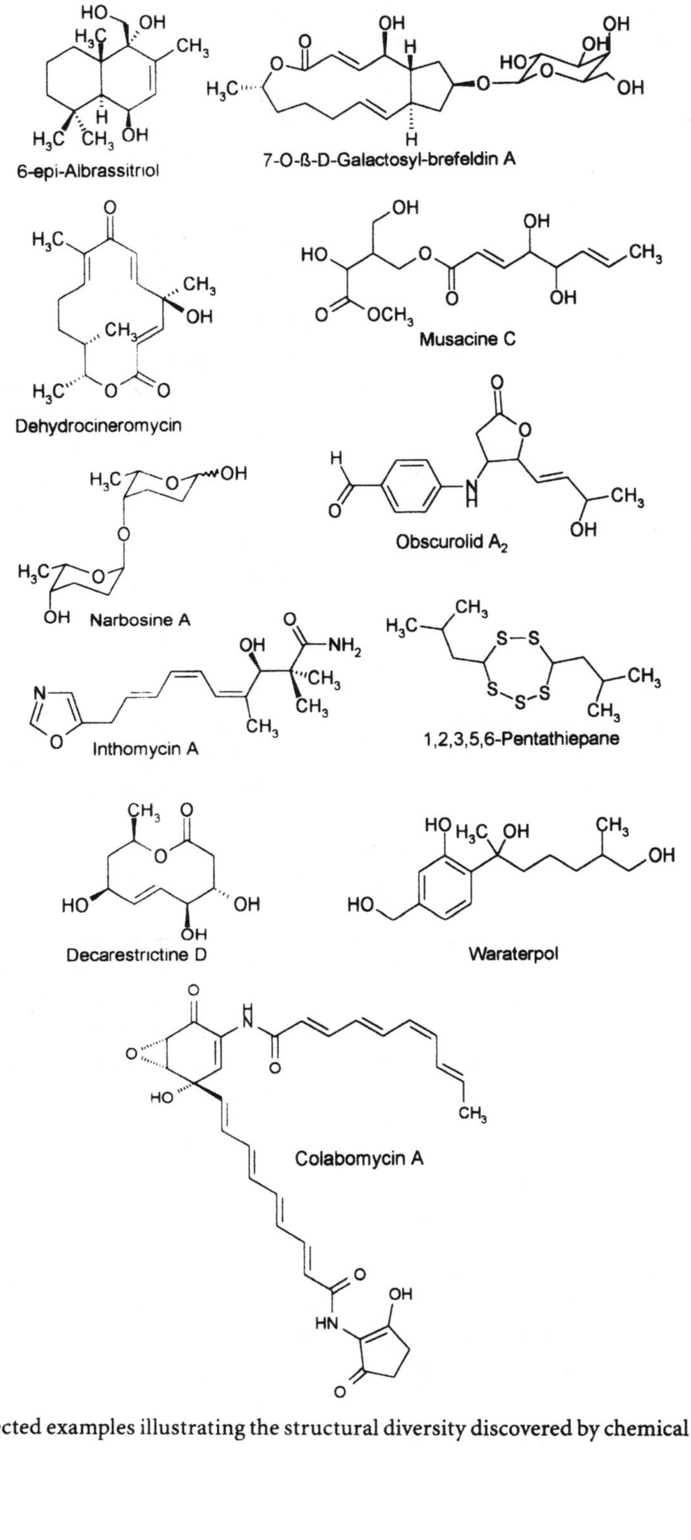

Fig. 8. Selected examples illustrating the structural diversity discovered by chemical screening

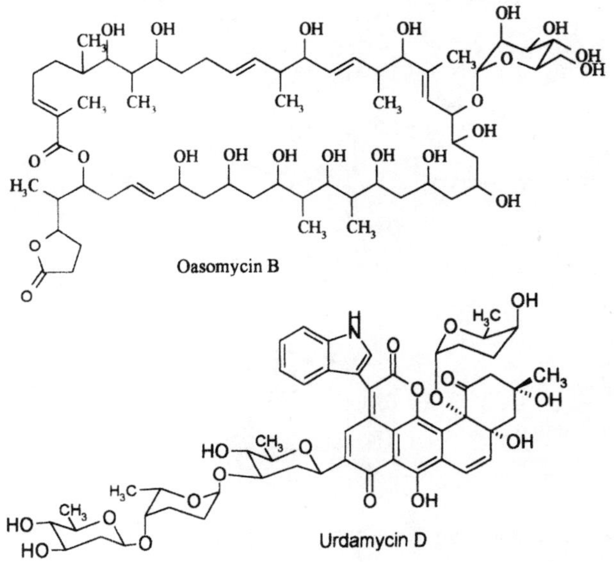

Fig. 8 (continued)

4.1.3
Future Potential

The future potential of the physico-chemical and the chemical screening approach lies in the possibility of accessing the outstanding structural resources from nature and to build up a collection of pure new and known natural products which can advantageously be used for broad biological screening. Natural compound collections of substantial structural diversity contribute to improved lead discovery, and efficiently support synthetic libraries (e.g. from classical or combinatorial synthesis). Often, it is better to run a biological screening with pure compounds from natural sources rather than with crude extracts. In order to extend the opportunities arising from testing pure compounds, in Germany a collection of natural products and derivatives is being established under the leadership of the HKI in Jena. The start up of the so called Natural Product Pool has been funded by both German companies and the German Government [194, 195].

4.2
Approaches to Enlarge Structural Diversity from Nature

Structure modification programs of biologically active compounds (leads) are an integral part of drug development. Among the many objectives of the derivatization of lead structures are gain of information on structure/activity relationships, enhancement of intrinsic potency, providing oral applicability, increase in bioavailability, elimination of unwanted side effects, and in case of antibiotics, overriding resistance mechanisms. Such lead optimization pro-

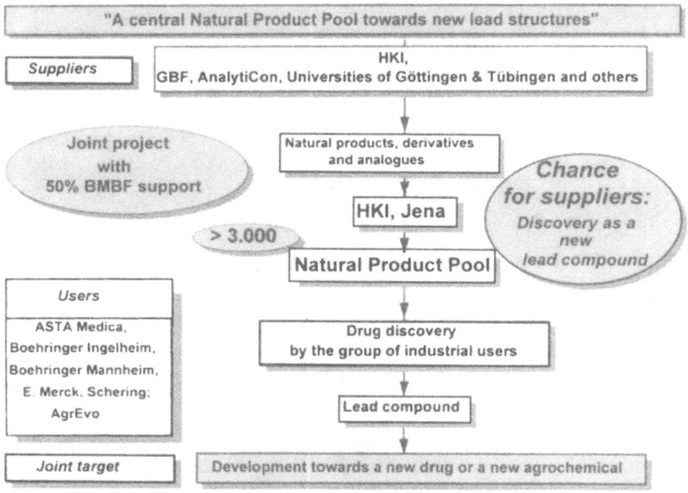

Fig. 9. The Natural Product Pool concept to identify new lead structures for drug and agrochemical development

grams are directed towards the selection of an optimal candidate for further drug development. In addition, modification of a given structure, either synthetic or from nature, may yield new profiles of biological activity, eventually giving access to a new lead compound. Furthermore, testing of compound libraries consisting of more than 100,000 molecules in the various target-directed bioassays designed for hit identification in HTS has shown that numerous molecules interact with more than one target. Therefore, enlarging chemical structure diversity in order to elaborate test sample libraries of the highest possible structural diversity is of superior interest for the primary screening run in HTS.

The chemical repertoire for structure generation includes derivatization and degradation of a given compound, as well as partial or total synthesis. Today, the well known "classical" strategies of organic and medicinal chemistry for the elaboration and variation of chemical structures are supplemented by new methods and technologies of "combinatorial chemistry". Combinatorial chemistry includes parallel synthesis, as well as split- and split-pool techniques [196–203]. On solid phase or in solution, combinatorial chemistry strategies allow efficient and systematic structural variation of a given chemical structure, and can advantageously be combined with structural studies on the pharmacological target. However, strategies of combinatorial chemistry can also be applied to design compound libraries for HTS aiming at lead discovery. Actually, combinatorial chemistry, typically in automated procedures, is of widespread use in pharmaceutical and agrochemical research and development. However, biological methods for structure modification can be a powerful supplementary tool for derivatization of structurally complex molecules from both, natural and synthetic sources.

4.2.1
Combinatorial Natural Product Chemistry

As a result of the limitations in test sample supply, combinatorial chemistry was developed which in consequence led to straightforward automation of processes necessary in organic chemistry. Applying solid-support peptide chemistry yielded multi-component systems comprising of more than 100,000 individuals. However, recent strategies aim at synthesizing medium-sized libraries consisting of more or less single components in each reaction vessel. This trend became necessary when complex mixtures derived from split-pool strategies often failed in HTS bioassays. In particular, limitations are substantial with respect to cell-based screening assays. At present, concepts based on solid phase synthesis still dominate synthesis in solution due to more facile product purification procedures.

In the past few years, most types of organic chemistry reactions have been transferred to solid-phase synthesis enabling the application of tools such as esterifications, acylations, synthesis of imines and oximes, substitution reactions, cycloadditions, olefinations, multiple component reactions, reductions, and palladium chemistry [197, 198, 204–206]. Because nature proved to be a most valuable source for structural diversity of low-molecular mass compounds, new concepts in order to supplement approaches in combinatorial chemistry by involving natural compounds as multi-functional building blocks attract increasing attention in pharmaceutical research. However, up to now only few examples have been reported in detail. Most strategies in combinatorial natural product chemistry focus on the total synthesis of natural products and its analogs. Mainly the synthesis of the alkaloid skeleton, such as indoles [207–209], 2-arylindoles [210–212], spiroindolines [213], 1,2,3,4-tetrahydro-β-carbolines [214–217], quinolines [218–220], and isoquinolines [221–223] is targeted. In addition, synthetic approaches aiming at the modification of e.g. prostaglandins [228, 229], balanol, a protein kinase C inhibitor from *Verticillium balanoides*, and epothilone (section 3.1) [224–226] have been reported. Combinatorial chemistry on the epothilone skeleton yielded a library of 112 derivatives synthesized on a modified Merrifield resin. Subsequent testing for tubulin assembling properties provided structure/activity relationships as outlined in Fig. 10.

Polyketides represent challenging targets for biomimetic combinatorial synthesis starting from simple building blocks. First attempts towards the preparation of libraries have been reported [230, 231]. However, at present combinatorial biosynthesis [232] seems to be superior in order to get access to "unnatural" natural products with a polyketide-type skeleton. Steroids are common templates for combinatorial approaches [233–236], but also studies with baccatin III [237], a precursor of paclitaxel and docetaxel, and some saponification products of alkaloids such as yohimbinic and rauwolscinic acid have already been reported [238]. In our own laboratories, we started developing methods towards automated solid-phase chemistry in parallel in order to decorate funtionalized natural product skeletons such as yohimbine, ergotamine, norscopolamine, nortropine, alkenyl substituted hydroxypiperidines, and terrecyclolic

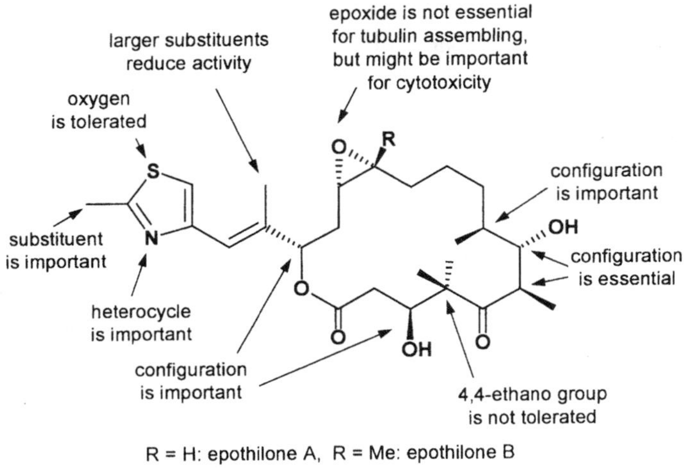

Fig. 10. Structure activity relationships of epothilone provided by combinatorial natural product chemistry

acid with various building blocks to achieve test compounds with molecular masses of more than 600 Dalton.

A new approach to obtain increased diversity in the search for new lead compounds called kombiNATURik has been co-developed by the German companies AnalytiCon and Jerini Bio Tools. The kombiNATURik program starts from natural compounds which are further diversified by solid-support chemistry introducing e.g. peptide or carbohydrate moieties. KombiNATURik libraries generally comprise several hundreds to thousands of single molecules derived from multi-parallel synthesis.

Future success in combinatorial chemistry will substantially depend on both, the structural diversity, and biological activity of compounds submitted to HTS. Up to now, structural diversity of combinatorial libraries is limited because their design is guided more by the chemical reactions currently adaptable to combinatorial chemistry, than by biological requirements. It turned out that low-molecular mass compounds of up to 400 to 500 Dalton if they are synthesized within conventional combinatorial chemistry attempts seem to fail affecting disease related protein/protein or protein/DNA interactions sufficiently. Combinatorial chemistry starting from natural compounds as building blocks or templates of approximately 400 to 500 Dalton resulting in molecules of 700 to 800 Dalton can fill this gap. These nature-derived molecules promise improved properties towards the inhibition of biomacromolecular interactions which play important roles e.g. in the regulation of cell growth and differentiation processes.

Today, it is obvious that in drug discovery combinatorial chemistry cannot displace natural product research. However, natural product chemistry can take advantage of combinatorial strategies. In our opinion, the discovery and development of innovative therapeutic agents will substantially benefit from the synergistic use of biological and structural-directed natural product screening approaches, natural product chemistry, and combinatorial strategies.

4.2.2
Biological Derivatization Methods

Since the early approaches on penicillin side chain modification [239–241], and 11α-hydroxylation of steroids [242, 243] which both mounted in important industrial processes, the techniques of biological structure modification have been further developed. Nowadays, a whole array of different methods using microbial materials and processes is ready for the derivatization of bioactive compounds. The various methods can be categorized into those employing the native biosynthetic machinery of a producing organism, those involving a manipulation of the biosynthetic pathways on the enzymatic or genetic level, and finally, the application of individual biosynthetic enzymes. The experimental demands on the application of the various methods range from "simple" feeding of biosynthetic precursors into standard cultivations to more sophisticated approaches involving genetic engineering of biosynthetic enzymes. Genetics are applied in the cell-based combination of biosynthetic genes from different strains or the in vitro reconstitution of biosynthetic pathways with overexpressed enzymes. A schematic overview of the different methods of biological derivatization pointing to the experimental procedures, the necessary biological material and the supply of precursors is provided in Fig. 11.

All of the various methods of biological derivatization become possible due to a relaxed substrate specificity of some of the biosynthetic enzymes, especial-

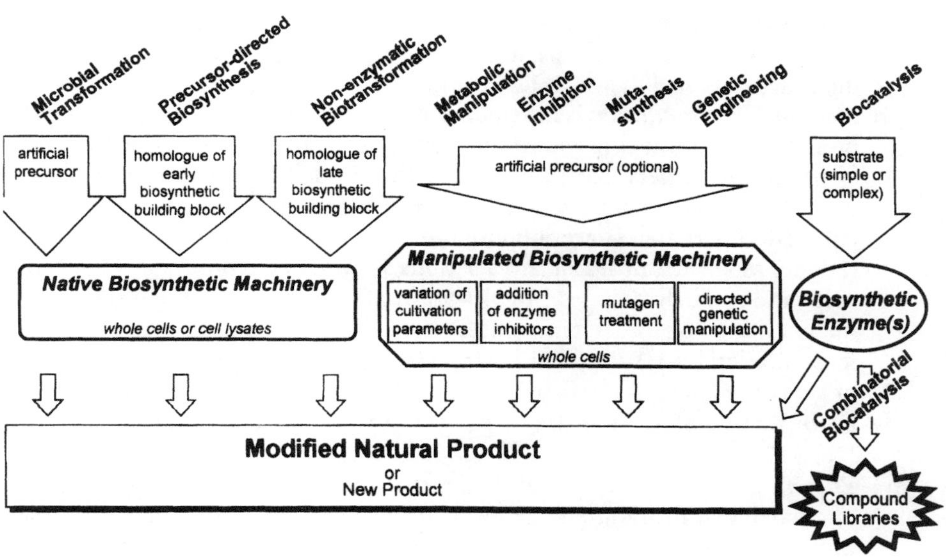

Fig. 11. Methods of biological derivatization

ly those of microbial secondary metabolism. Of the many well-known advantages of the use of enzymes as catalysts, the high chemo-, regio- and stereoselectivity, as well as mild reaction conditions are particularily advantageous when used with natural products. Comments on the most wide spread methods of biological derivatization which are based mainly on the use of microorganisms are summarized below:

i) *Microbial transformation* makes use of enzyme catalyzed reactions with living cells, typically exploiting a single chemical reaction, like oxidation, reduction, hydrolysis, acylation, phosphorylation, glycosylation, methylation, amination, halogenation, isomerization or formation of N-oxides [244–246]. The methods of application, as well as the large variety of types of microbial transformation reactions are subject of some excellent review articles [242, 245–256]. There exist examples of highly complex molecules where microbial oxygenation has been shown, e.g. hydroxylation reactions of the macrolide antibiotics avermectin and rapamycin [257–259], hydroxylation of the polyether grisorixin [260], and carbonyl reduction of midecamycin A_3 [261]. In general, experiments of microbial transformation are easy to handle. However, the prerequisite is a screening of a large number of organisms for a desired enzyme activity.
ii) *In precursor-directed biosynthesis* the derivatization of a secondary metabolite is achieved by adding analogues of a biosynthetic precursor to the cultivation of a producing organism [262]. Thus, organisms capable of incorporating artificial precursors into their enzymatic processes provide modified metabolites. Precursor-directed biosynthesis requires knowledge about the biosynthesis of the original product, especially the basic building blocks and major biosynthetic intermediates. The method has been successfully introduced to industrial antibiotic production of penicillins, aminoglycoside antibiotics and avermectins, among others. Most examples of precursor-directed biosynthesis, however, are of exploratory nature, and products are obtained in lab-scale amounts.
iii) *Metabolic manipulation* is based on the variation of cultivation parameters of antibiotic producing microorganisms. Such manipulations of the biosynthetic machinery have resulted in the production of new metabolite derivatives, presumably involving post-genetic alterations. The variations of cultivation parameters are supposed to be drastic changes ("stress fermentation") rather than parameter variations as in optimization experiments of cultivations. Such changes in cultivation parameters that might influence metabolite patterns are temperature as in heat shock or low temperature fermentation [263]), mechanical stress (e.g. changes in rheology or addition of glass beads), light [264], pH-value, addition of ingredients to the culture broth [265, 266], or increase of oxygen partial pressure up to 1,200 mbar [267–272].
iv) *Enzyme inhibition* as a biological derivatization strategy works by blockage of a particular biosynthetic pathway through administration of an enzyme inhibitor to the growing culture. Generally, this method can yield three different kinds of products: biosynthetic intermediates which are accumu-

lated because of an interruption of the biosynthetic pathway, derivatives of this intermediate formed through non-enzymatic reactions or enzymes that are part of the biosynthetic sequence behind the interrupted step, and derivatives of added artificial intermediate homologues. The most widely applied inhibitor for biosynthetic enzymes of secondary metabolites is cerulenin which efficiently inhibits polyketide, fatty acid and steroid biosynthesis [273–278]. In a number of cases, cerulenin has been used to generate structure variants of macrolide antibiotics by inhibiting the native macrolide biosynthesis. Furthermore, its allows the exploitation of late biosynthetic steps, especially glycosylation reactions by making use of added aglycon analogues [279–282].

v) The concept of *mutasynthesis* [239, 283–286] is based on the random mutation of a biosynthetic sequence aiming at the interuption of a single enzymatic step. Thus, mutasynthesis can yield new derivatives of a secondary metabolite similar to the application of enzyme inhibitors. The necessary mutants of a producing organism can either be identified from spontaneous mutations, or more efficiently through induced mutagenesis by mutagenic chemicals or UV-irradiation. For decades, mutasynthesis has been a common strategy for the production of derivatives of secondary metabolites, as well as for studies of biosynthetic pathways. Successful applications of the mutasynthetic method have been described for macrolides [287, 288], anthracyclines [289–295], polyethers [296], aminoglycosides [297–299], ansamycines [300–302], and nucleoside peptide antibiotics [303–305]. The production of the staurosporine aglycon K-252c [306], new avermectins [307], or several new angucyclinones [308, 309] are recent examples of mutasynthetic approaches.

vi) The production of *hybrid antibiotics via genetic engineering* of biosynthetic pathways is one of the major developments in natural products research [310, 311]. Especially the combination of biosynthetic genes from different organisms in one host organism, termed "*combinatorial biosynthesis*", gave access to new compounds embodying features from different metabolites. A future potential will lie in the combination of different biosynthetic pathways, e.g. combining the enzymes of polyketide metabolism with recently identified enzymes of non-ribosomal peptide formation or carbohydrate biosynthesis. The particular potential of genetic engineering as method of biological derivatization has spurred the engagement of a number of pharmaceutical companies in this area of natural products research and a number of newly started companies are based on this methodology. After the first genetically engineered hybrid antibiotics had been reported in the early eighties (mederhodin A/B and dihydrogranatirhodin [312]), lots of research efforts in the past four years resulted in a number of hybrid metabolites [313–315].

vii) The *use of individual enzymes* for biological derivatizations, either as crude preparations from genuine sources, or in pure form from overexpression systems, is the modern counterpart of cell-based methods like microbial transformation and precursor-directed biosynthesis. So far, numerous enzyme catalyzed reactions have been reported [316, 317]. Especially in carbohydrate chemistry this concept of reconstituting a pathway or a combination of interlinked pathways has found many interesting applications [318, 319]. A new and powerful application for enzymes as catalysts in synthetic chemistry are combinatorial strategies for the generation of whole libraries of organic compounds [320, 321]. Enzymes are able to catalyze specific conversions of mixtures of complex molecules in solution under mild conditions and without by-products. Up to now, the applicability of *combinatorial biocatalysis* has been shown on an array of structurally diverse organic molecules. A library of nearly 200 analogs allowed to identify acylated derivatives of paclitaxel that overcome solubility problems by a 58-fold (paclitaxel 2′-adipoylglucose) and a 1625-fold (paclitaxel 2′-adipic acid) increase in relative water solubility [322]. As an example the derivatization program performed on the natural product bergenin by a two-staged protocol is depicted (Fig. 12) [320, 321].

5 Developments in Drug Discovery Technologies

5.1 Potential Impact of Genomic Sciences

All drugs that are currently on the market – excluding antiinfectives – can be put down to 417 different disease related human targets comprising primarily G-protein-coupled receptors, enzymes, hormones, growth factors, ion channels, and other biomacromolecules [3]. It is obvious that today most diseases cannot be cured. This may implicate that the corresponding disease targets so far known give access to good drugs only in a few cases. However, some targets such as the angiotensin-converting enzyme or the adrenergic receptor subtypes may be close to ideal targets, as drugs affecting these targets represent highly effective and safe antihypertensive therapeutics [3].

On the other hand, about 33% of the today's drug targets are related to neurological disorders without realizing satisfying therapeutic approaches to any of the serious neurodegenerative diseases. The conclusion drawn in pharmaceutical industry from this fact is quite remarkable. It is not assumed that further screening of compounds could result in an optimized drug. On the contrary, it is expected that improved therapy is only accessible by identifying and affecting new targets. These new disease related targets can be approached by structural and functional analysis of the human genome. Therefore, genomic sciences actually have become a key issue in order to accelerate drug discovery and development in future.

Currently, research in the pharmaceutical industry is concentrated on about 100 diseases that in their opinion really matter. In order to calculate how many targets may refer to these diseases of interest the following is estimated from diseases with well understood pathways such as diabetes type II and hypertension [3]: Most diseases are caused by genetic factors, and proved to be multifactorial. The number of genes involved to contribute to the various disease phenotypes turned out to be around five to ten, thus resulting in 500 to 1,000

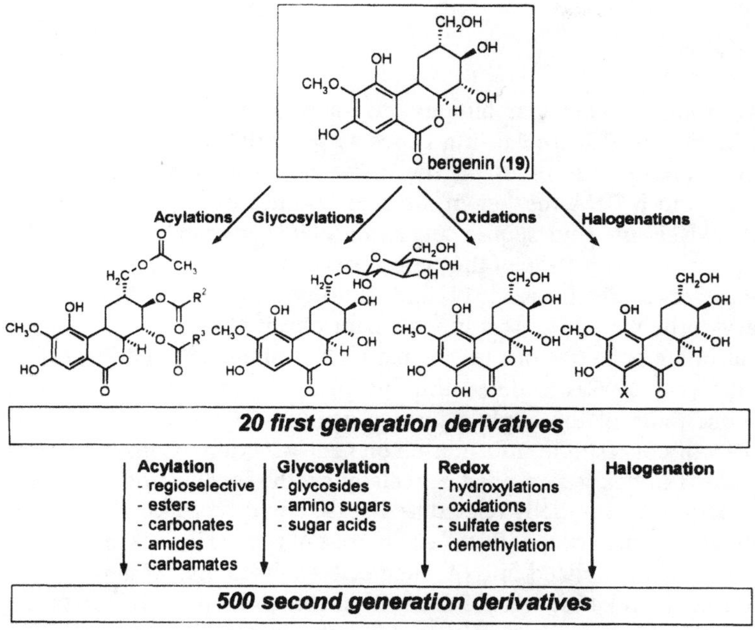

Fig. 12. Derivatization of bergenin as an example for the concept of combinatorial biocatalysis

disease related genes and gene products. Further estimating that each of these disease related genes or their products interact with three to ten proteins in the signaling pathways at least yield 3,000 to 10,000 genes or gene products that are worth to be tested as potential targets for interaction with effectors of low-molecular mass.

The identification of disease related targets is one of the most challenging issues in today's drug research process. Accelerated information on gene structures is provided by public and private genomic databases providing raw DNA sequence information on the human genome as well as on the genome of various animals, plants and microorganisms. Gene sequences from pathogenic bacteria and fungi attract considerable attention. New targets are urgently requested to treat infections caused by both, antibiotic resistant microbial strains and novel types of bacteria such as *Helicobacter pylori*. Actually, the number of disease-causing bacteria whose genome is completely sequenced by routine DNA sequencing techniques combined with adjusted data management systems is growing rapidly.

Traditionally, the approach for the identification of biological targets was a reductionist one. The pathological phenomenon was examined with increasing resolution typically starting from an understanding of fundamental biological mechanisms in man or in an animal model. These examinations were followed by studies involving intact tissues and cells or preparations thereof, ultimately leading to the identification of molecular targets for drug interaction [323].

5.1.1
Analytical Methods

The revolution in molecular biology gave access to new techniques enabling the elucidation of disease-causing genes as potential new targets for therapy. Positional cloning, e.g. relies on the discovery of an association between a phenotype and a DNA marker in affected families or animals [323]. By subsequent analysis and gene sequencing a candidate gene can be revealed, as it was shown for the identification of the genetic defects underlying the human monogenic disorders cystic fibrosis [324] and Huntington's disease [325], or the recent discovery of the murine gene responsible for obesity [326].

The ultimate objective of the international Human Genome Project scheduled for the year 2005 is to determine the DNA sequence of the approximately 3×10^9 base pairs of the haploid human genome. However, reading the DNA sequence will not result in information on gene functions. Consequently, a great deal of effort is being expended especially on the localization and functional analysis of genes potentially related to disease phenotypes [323].

Expressed sequence tags (ESTs) e.g. represent a special class of DNA markers [327] derived from spliced mRNA. Such cDNA copies of mRNA can be generated from cells and tissues by PCR (polymerase chain reaction) techniques. Transfer into the bacterium *Escherichia coli* gives access to a cDNA library consisting of more than 10^6 clones that represent the entire repertoire of genes expressed in the tissue of interest at the time of isolation. The nucleotide sequence can be determined for short regions of individual cDNAs, and this provides a unique identifier, or EST, for that cDNA representing a specific gene. At institutions such as Human Genome Sciences (HGS), The Institute for Genomic Research (TIGR), Incyte, and Washington University together with Merck are applying the EST strategy coupled to powerful computation to robotic high-throughput technologies towards the determination of the nucleotide sequences of thousands of clones within days. However, an EST by itself frequently cannot be ascribed a precise function, but a novel DNA sequence can be analyzed by comparison of its nucleotide sequence within a databank of known sequences. In case of homology with known superfamilies of genes, such as protein kinases, cytokines, or G-protein-coupled 7-membrane-spanning receptors, a related function can be assigned [323].

Alternative techniques such as subtractive hybridization and differential display cloning [328], however, lead directly to the identification of genes that underly phenotypes. These techniques focus exclusively on genes that are differentially expressed in different cell or tissue samples. Differential display cloning permits a simultaneous comparison of genes that are either up-regulated or down-regulated within different mRNA samples of interest which for instance derived from cells or tissues of healthy people and patients suffering on certain disease phenotypes. Consequently, by applying the strategy of differential display cloning new genes probably related to a disease phenotype of interest can be identified directly, thus leading to a new potential drug target [323].

In order to accelerate the process of target identification and target selection prior to bioassay development and high-throughput screening, chip hybridization techniques have come into use. Microarray chips permit the simultaneous analysis of thousands of genes and their expression profiles in different cells and tissue types, and under specific conditions, thus contributing to accelerated target elucidation. Furthermore, the microarray chip technology can be applied for improved gene diagnostics.

5.1.2
Applicational Aspects of Gene Function Analysis

However, the discovery of a new potential drug target represents only the first step in a series of experiments towards bioassay construction and integration into the HTS process. Valid cell culture experiments, and qualified transgenic animal models, respectively, are prerequisites for constructing a predictive screening assay. Preliminary studies of the function and regulation of so-far unknown human genes can be performed by well characterized model orga-

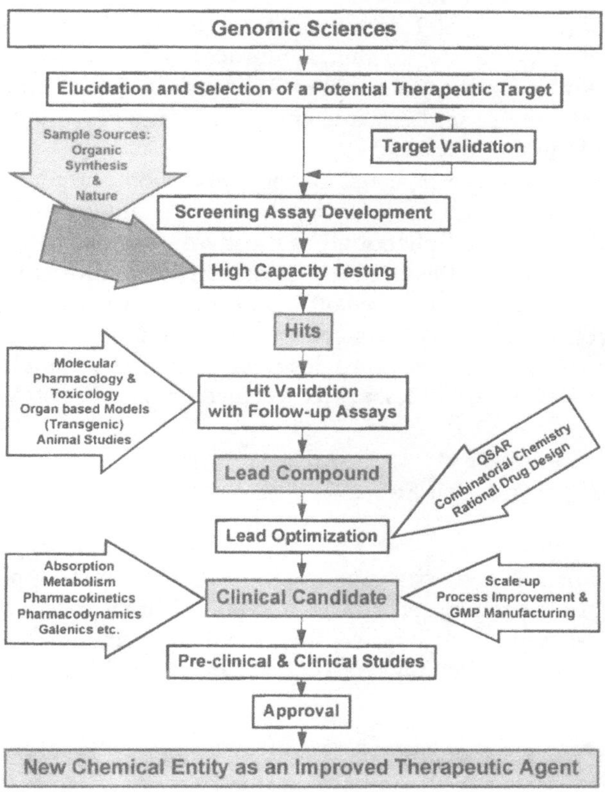

Fig. 13. The drug discovery and development process towards a new therapeutic agent starting from genomic sciences and target selection

nisms such as Yeast or Drosophila. Gene functions related to cell proliferation can favorably be studied in *Saccharomyces cerevisiae* whilest *Drosophila melanogaster* is preferred for examining signaling pathways [329]. DNA sequence analysis of *Saccharomyces cerevisiae* has been completed recently, and gene function analysis is rapidly progressing, thus contributing to facilitated function assignment of various novel human DNA sequences and genes.

In fact, the precise function of numerous human genes can only be studied in a mammalian model. From the beginning of animal studies in the industrial drug development process the mammal of choice has been the mouse. Homologous proteins between mouse and man typically share over 90% of their amino acid sequences. Physiology, embryology and genetics of the mouse have been well characterized over many years. Referring to human genomic sciences, one of the most important technologies developed so far is the generation of transgenic mice by introducing any genetic change that can be transferred into a live mouse and bred into subsequent generations [329]. However, the basic techniques for target validation are accessible in principle, but it takes a long time and a resources-consuming process to prove the impact of affecting a selected target gene on a certain disease phenotype. Consequently, it is actually speculated that target validation can be omitted. Rapid bioassay development and transfer to the HTS process promise the identification of appropriate binding low-molecular mass ligands in a short time. These ligands are potential drug candidates, and thus can be used favorably to exploit the relevance of the selected target.

Genomic sciences are expected to identify accessible targets for specific drug interaction. However, the majority of diseases are multifactorial, with the consequence that a defined phenotype is mostly due to a specific genotype. This indicates that in future the therapeutics market has to diversify substantially to provide tailor-made drugs relevant to the respective disease genotype. This seems to contradict today's strategies of the global operating pharmaceutical companies aiming at drugs with sales rates of at least several hundred million US $ per year which are necessary for a return on their investment.

5.2 High-Throughput Screening Assays

Before the advent of molecular biology and gene technology, whole-cell and whole-organism assays with non-specific endpoints were applied to discover and evaluate bioactive compounds. Today, a target-based approach using mechanism-based screens has replaced the non-specific assays. These target-directed bioassays following the key-lock principle fall into two major categories:

i) receptor- or enzyme-based screens using a particular target of interest in a purified system, and
ii) cell-based assays using engineered eukaryotic cells or microorganisms.

In the latter systems, introduction of reporter systems based on the transcription of genes encoding proteins such as β-galactosidase, luciferase, alkaline phosphatase or, more recently, the green fluorescent protein from jellyfish give rise to monitor the target of interest [330, 331].

In order to realize a high throughput of test samples, automation and miniaturization of the assays as well as an appropriate computing power and data management are required. Reproducible screening results depend substantially on assay reoptimization accommodated to the type of equipment employed. Furthermore, the throughput of the assays depends upon the complexity of the linked steps and the read-out times, respectively. Biochemical assays based on ligand binding to a receptor, or on enzyme catalyzed turnover rates will generally yield higher throughput than cell-based assays which need assay times of approximately a few hours to several days depending on the type of target processed, and the type of cell applied. Today, an ELISA or scintillation proximity assay (SPA) performs the highest throughput because these assays require relatively little manipulation. Furthermore, these types of assay give very reproducible results with a high signal-to-noise ratio when the binding constant is in the low micromolar range or less [332]. Target enzymes such as phospholipases, kinases, phosphatases, and proteinases can favorably be adopted to HTS procedures. Thus, inhibitors and activators of the enzyme of interest can be easily identified.

Biochemical screening assays based on receptor binding analysis exclusively rely on cell-free interaction systems. They do not provide any information on the type of interaction, that means whether an affecting test sample acts as an agonist, or an antagonist of the endogenous ligand of interest. In the past, numerous valuable drugs have been identified by applying in vitro binding assays. However, trends to more complex cell-based screening assay systems that give information on the function of bioactive principles in the living cell are progressing [330, 331, 333]. This trend is obviously due to results from cell biology on the impact of the complex network of specific biomacromolecule interactions on regulatory mechanisms of the cell. Alterations in the respective biomacromolecules such as proteins and DNA often lead to unregulary interaction properties causing abnormal gene expression rates that are directly related to various phenotypes of human disorders. Therefore, therapeutic approaches currently are expected from drugs that specifically affect these interactions. In order to realize predictive screening assays cell-based systems are often preferred.

Mammalian cells are expensive to culture and difficult to propagate in automated systems. An alternative is to recapitulate the desired human physiological process in a microorganism, such as yeast [330, 331, 334]. For instance, signaling through human G-protein-coupled receptors has been reconstituted in *Saccharomyces cerevisiae* to yield a facile growth response or a reporter gene read-out. Similarily, mammalian ion channel receptors can be reconstituted in yeast to realize a readily assayed growth response. Furthermore, protein-protein interactions as well as peptide hormone receptor binding have been faithfully reproduced in yeast using the two-hybrid system [334]. Variations of the two-hybrid system give rise to the one-hybrid, and the tri-hybrid (tribrid) system [335], respectively. Finally, it was shown that many mammalian transcription factors operate in yeast [330], these include steroid hormone receptors such as the glucocorticoid [336], the estrogen [337–340], the progesterone receptor [341 – 343] and the vitamin D receptor [344, 345].

Easy handling, short generation times, ready genetic manipulation, continuous heterologous gene expression, resistance to solvents and low cost of growing make yeast an attractive option for cell-based HTS attempts. However, screening assays in *Saccharomyces cerevisiae* can only serve as primary assays that have to be followed up by advanced profiling of the "hit" in tissue culture systems, organ systems and animal models, respectively, before lead structure commitment.

With the cell-based screening approach, a recombinant cell or microorganism is engineered to respond in a specific manner to an effector of the molecular target of interest permitting the examination of complex multiprotein interactions without having to reconstitute the entire system in vitro. If the system is properly formatted, the probability of a lead compound derived from a cell-based screening being active in pharmacological models is supposed to be greater than if the compound was derived from a cell-free assay using purified macromolecules. Furthermore, the need for the agent to cross the cell membrane to demonstrate activity in a cell-based assay probably eliminates many nonspecific false positives obtained from screening attempts using isolated enzymes and receptors [333]. For many targets, cell penetration is required for pharmacological activity. On the other hand, potential lead compounds that inhibit the target but cannot cross the membrane will not be discovered. These leads may still be useful as subsequent chemical modification could introduce cell penetration properties. Therefore, the use of cells with permeability mutations provides a good compromise, eliminating many false positives while allowing a variety of potential lead compounds to be detected.

A major problem of cell-based screening assays is that cytotoxicity derived from mechanisms unrelated to the target of interest prevents further evaluation of a compound or, more feasible, of an extract from natural sources. Furthermore, depending on the assay endpoint, false positives may be selected. Therefore, secondary assays are required to eliminate cytotoxic false positives as well as to prove the mode of interaction that gave positive endpoint detection in the screening assay.

In the next few years, quantitative changes in both genetic and instrumentation engineering, including advances in nano-technology, will affect cell-based HTS approaches. New reporter systems resulting in sensitive fluorescent readouts without cell disruption will probably contribute to assay miniaturization and faster screening runs. The construction of assays that take advantage of the ability of cells to grow and to proliferate will cause a qualitative shift in how HTS is conducted. Application of such positive selections where cell growth depends on the addition of the bioactive principle of interest should allow examination of hundreds-of-thousands to millions of test samples per day, rather than the tens-of-thousands now accessible using the best screening formats [330].

Furthermore, novel methods for single molecule detection will contribute to improved drug discovery. For instance, fluorescence correlation spectroscopy (FCS) is characterized by the ability to measure single molecules in sub-microliter sample volumes with a typical range of sensitivity between 10^{-6} and

10^{-12} mol l^{-1} for homogeneous assays [346]. FCS can be used for both, studies on molecular interactions in solution based systems, or in cells performing assay times in the second range. The application of FCS in accelerated HTS processes is currently demonstrated by the German start-up company Evotec.

5.3
Sample Supply in High-Throughput Screening

Regarding current capacities of primary screening assays realized in HTS attempts, it has become evident that high sample numbers to feed the assays are needed. Combinatorial chemistry and the exploitation of molecular diversity derived from biodiversity currently offer the greatest promise of finding novel lead structures for significant therapeutic areas [4, 347, 348]. Especially new chemical entities (NCEs) are required in order to facilitate therapeutic innovations. Today, in the pharmaceutical industry, chemical libraries consisting of at least 100,000 single compounds and libraries generated by combinatorial chemistry approaches are routinely submitted to HTS in order to identify new lead structures valuable for drug development. At the moment, crude extracts derived from natural sources, such as plants, fungi, bacteria, and marine organisms only play a minor role.

5.4
Automation to Accelerate Drug Discovery

The starting point for the development of automation concepts was in analytical routine laboratories (e.g. analytical R&D and quality control) dealing with high numbers of samples routinely to be analyzed by the application of standardized operation procedures [349]. In parallel, the first robot systems for research were developed and these combine the key technology of liquid handling with the movement of items through a defined space.

Up to the present, the tremendous development in soft- and hardware as well as the needs of the customers in the rapid growing environment of drug discovery have resulted in a kind of evolution in purchase, installation, and implementation of laboratory robotics. Reliability, reproducibility, flexibility, and speed have been set to a high standard. Over the past 15 years, laboratory automation has grown from a novel technology to a powerful and easy-to-handle tool and provides the basis for modern high-throughput drug screening approaches.

5.4.1
Economic Aspects

Optimization of drug discovery strategies is driven by the actual patent situation (drug discovery and development 10 to 12 years; term of patent 20 years; expiry of important patents in the next few years will cause drastic drops in turnover) forcing the companies to shorten the time taken to bring a drug onto the market.

In addition, a more efficient primary drug screening and hit validation is necessary to minimize the costs. In total, for the discovery, development and the introduction of a new drug to the market including the costs for discontinued development projects about $300 to $500 million have to be estimated [350]. On the other hand, the number of new chemical entities (NCE, ca. 50 per year) is nearly constant since 1983 while the costs increased more than three times [351].

On this background laboratory automation in the early stages of drug development is able to enhance productivity, efficiency, reliability, and speed without increasing research costs. Typical assay costs in HTS are 1$/sample. Testing 200,000 compounds in 20 screens would then cost $4 million. It is expected that the costs can be brought down to less than $0.1/sample in defined cases by making use of effective screening technologies and miniaturization. However, sample sourcing from automated parallel synthesis or extracts from natural sources is even more expensive [352]. HTS therefore is not an inexpensive enterprise.

5.4.2
Sample Sourcing

Natural products and its analogs are important sources for novel lead structures. Due to the complexity of cellular metabolism, extracts from natural sources usually contain numerous different components in varying amounts. Therefore, integration into automated drug screening approaches adds additional efforts. Dealing with rough or enriched extracts from natural sources is highly cost-intensive but of remarkable interest. Up to the present, extraction procedures in order to prepare samples for drug screening are usually performed with a low automation rate. There exist a strong need in automation approaches for routinely performed procedures.

Recently, Merck KGaA in Darmstadt in cooperation with AnalytiCon AG (Berlin) presented a HPLC-based Workstation (SepBox) designed for the extraction and separation of plant material in a more preparative scale which has also successfully been used for the separation of secondary metabolites from culture broths of microorganisms. On the other hand, it has been shown that automated solid phase extraction (SPE) can achieve high quality samples from cultivation of microorganisms in an easy and cost-effective manner. In the latter case modified Zymark RapidTrace modules have been used in the automation concept which advantageously does not need HPLC-techniques [353].

Analysis of the sample quality either from synthesis approaches or from natural sources is usually performed by techniques like HPLC and capillary electrophoresis (CE) coupled to UV/VIS-, MS-, IR-, or NMR detection. Because the working procedures of these instruments are more or less serial, there exist an enormous need in new technologies allowing parallel performances and further miniaturization.

5.4.3
Sample Handling

While liquid handling in volumes >5 µl is reliable, procedures like sample storage and retrieval, weighing, dissolving, and distributing are more sophisticated. In principal, sample storage is performed following two different strategies: in diluted form [e.g. in dimethylsulfoxide (DMSO)], or as pure samples either in tubes or in deep-well plates. Besides defined storage conditions which should minimize stability problems, a reliable and fast retrieval of samples is required. Automated systems should be able to handle hundreds of thousands of different compounds [storage, multiplication, sample (back-) tracking, movement, etc.] with efficient logistics. Due to the different needs the automated storage and retrieval concepts installed in pharmaceutical companies are more or less customer designed.

Automated sample preparation suitable for the assays starting with solid or oily material is rather complicated and causes problems. This highly labour intensive procedure caused problems when it was done manally due to the nature of the material to be handled. Thus, only certain percentages of the compounds delivered in solid form can be handled by robotics. For this procedure in screening companies, only a low rate of automation is observable so far. Nevertheless, first automated filling stations exist in which a defined amount of a powder can be transferred from standardized vessels to test tubes for subsequent dilution.

Due to the fact that in most companies the sample storage and retrieval facilities are centralized, the test samples have to be transferred to the screening groups at various locations. Therefore, logistics such as bar-coding, data transfer, and delivery are crucial points. Delivery of the samples is performed usually in 96-well format after sealing with removable films in solution (e.g. on dry ice in DMSO), or as neat-films. Workstations able to seal microplates, label with bar-codes, and remove organic solvents in order to prepare neat-films are commercially available.

5.4.4
Screening Systems

Automation of assays based on nearly every biochemical effect like enzymatic reactions, cell-surface receptor-, and intracellular receptor binding, protein-protein-, and protein-nucleic acid interaction, cell adhesion, etc. has been reported. Robotic systems used in biological screening can be devided into workstation-type systems [354] and integrated systems [355]. Integrated systems are custom-made and allow one to integrate nearly every peripheral module necessary for screening courses.

As a recently developed technology, a novel automation concept has been presented which is based on a mover, a stacking device, and a reader or washer (Twister, Zymark). In comparison to the above mentioned systems, assays based on the Twister concept are only semi-automated. The outstanding benefit is the simplification of feeding, handling, set-up, as well as the small laboratory space needed. The capacity can be up to 20 to 80 microplates while higher throughput is achievable by making use of multiple units.

Although laboratory robotics is a well-established technology there exists a need for further improvement and next-generation devices, the so-called ultra-high-throughput screening (UHTS). The whole process of drug discovery is being forced to test an even higher number of compunds and a faster turnover of new assay systems. Starting in 1985, 10,000 data points per target have been produced per year. In 1990, the same amount was tested in a month, while in 1995 this was achieved in a period as short as a week. At present, there is a need to generate 10,000 to 100,000 data points in one robotic system per day. To meet these requirements, two approaches have actually been realized, the so-called Allegro system from Zymark (with multiple plate movers), and systems which revolve around microliter volume assays in 1,536-well microplates (Pharmacopeia) [356].

5.4.5
Data-Handling

Data management in drug discovery is used to handle the acquisition, analysis, report generation, and database functions of the drug discovery data. It is responsible for compound, and microplate registration, compound tracking, as well as inventory control [357]. The data handling is also used for the coordination of sample generation (combinatorial or parallel synthesis [358], natural product extraction etc.), sample preparation (mother-, and daughter plates etc.), assay registration, microplate schedule, data recording from analytical instruments (readers etc.), and data analysis. Finally, it is necessary to link biological results with chemical structures in order to complete the structure/activity relationship (SAR) feedback loop [359].

A key component to the success of any drug discovery program is the integration of HTS and combinatorial chemistry with database management and molecular diversity systems to yield an efficient and cost-effective method of organizing chemical information and biological data. The ongoing increase in throughput has created a substantial data management bottleneck. Furthermore, there are no standards for vendors for control software or file formats. This leaves a gap between the laboratory equipment and the database that can only be filled through custom programming. In comparison to the whole high-throughput screening process, the set-up and management of the data handling is cost-, and personal intensive.

6
Discussions and Conclusions

6.1
Natural Products

Currently, natural products are passing through a phase of reduced interest in drug discovery because of the enormous efforts which are necessary to isolate the active principles and to elucidate their structures. However, if one considers the diversity of chemical structures found in nature with the narrow spectrum of structural variation of even the largest combinatorial library it can be expected that natural products will become important again [4]. Mainly actinomycetes, fungi and higher plants have been shown to biosynthesize secondary metabolites of obviously unlimited structural diversity that can further be enlarged by structure modification applying strategies of combinatorial chemistry. Probably, a variety of novel concepts in natural products research is required to attract interest in incorporating natural sources into the HTS process.

For instance, a "Natural Products Pool" comprising only pure and structurally defined compounds has currently been established in the 96-well microplate format in Germany. The project aims at providing an infrastructure that improves the availability of natural products, their derivatives and analogs for industrial screening assays restricted to the minimum amount of test sample needed. Only a minority of the natural products known so far have been biologically characterized in detail. Therefore, any novel target-directed screening assay may result in identification of a new lead structure even from sample collections comprising already described compounds.

Recent strategies in combinatorial chemistry aim at synthesizing single compounds or definite mixtures of only a few components. This trend was accelerated when it became evident that complex mixtures derived from "split and combine" strategies often failed in HTS bioassays. Limitations are substantial with respect to cell-based screening assays. These facts indicate that results from testing traditional extracts from natural sources such as plants or microbial cultures have to be questioned. Therefore, generating high quality test samples of less complexity and accelerated reproducibility is favorable in order to compete successfully with combinatorial chemistry.

Today, selected targets of interest are transferred to the HTS aiming at discovering a "hit", and subsequently a lead structure within one to two months. After that time, the target concerned is replaced by a new one. Thus, rapid characterization and structure elucidation of the active principles of interest from natural sources are critical with reference to competition with synthetic libraries of pure and structurally defined compounds. Elaboration of LC-MS and LC-NMR techniques to accelerate structure elucidation of bioactive principles from natural sources is currently underway. Combination of these techniques with databases comprising a maximum of known natural compounds will probably contribute substantially to more interest in natural products for application in drug discovery.

Indications that today only a small percentage of living organisms have been described implies that there is an enormous reservoir of natural compounds of great structural diversity which are still to be discovered. The United Nations Convention on Biological Diversity adopted in Rio de Janeiro in 1992 sets the basic principles of access to and exploitation of global biological sources in the future. The Convention introduces national ownership of biological resources. As a result, various pharmaceutical companies fear that this will cut off the flow of genetic resource materials for industrial drug discovery. However, the Convention has not yet been ratified by the USA.

For increasing effectivity and decreasing costs for HTS technologies, the basic limiting factor for finding new lead compounds will be the supply of structural diversity. The relevance of natural products in drug discovery and development will consequently depend to a great extent on the efficiency and costs of accesss to chemically diverse extracts from natural sources compared to the supply from synthetic sources [360].

6.2 Technologies

Currently, dramatic changes in drug discovery technologies are taking place affecting any issue that plays a role in the lead-finding process. In future, target elucidation will probably derive mainly from genomic sciences, although lines of argumentation are arising that stress the relevance of target gene finding starting from given bioactive agents performing valuable in vivo activity. Progress in lead discovery is expected to depend substantially on methods and techniques that accelerate validation of new target genes potentially related to human disorders. Validation can be provided either by functional gene expression in valid animal models, or by immediate transfer to HTS in order to get rapid access to a low-molecular mass effector that gives rise to both a drug candidate and target validation.

Today, limitations in HTS are not only due to missing high quality test sample numbers but also due to quantitative limitations in the availability of target reagents. In order to compensate for the shortcomings in target supply further progress mainly in heterologous gene expression is required, either towards advanced cell-based assays or towards target isolation and application in a defined molecular system.

Acknowledgement. The authors would like to thank Dr. Kai U. Bindseil (AnalytiCon AG, Berlin) for valuable help for Sect. 4.2.1 "Combinatorial Natural Product Chemistry", and Dr. Isabel Sattler (HKI, Jena) for Sect. 4.2.2 "Biological Derivatization Methods".

7 References

1. Cragg GM, Newman DJ, Snader KM (1997) J Nat Prod 60:52
2. Müller H (1998) Oral Presentation, 10. Irseer Naturstofftage, Germany
3. Drews J (1996) Nature Biotechnol 14:1516
4. Kubinyi H (1995) Pharmazie 50:647
5. Hoagland RE (1988) Naturally occuring carbon-phosphorous compounds as herbicides. In Cutler HG (ed) Biologically active natural products: Potential use in agriculture. ACS Symposium Series 380, Washington, DC, p 182
6. Bayer E, Gugel KH, Hägele K, Hagenmaier H, Jessinow S, König WA, Zähner H (1972) Helv Chim Acta 55:224
7. a) Kondo Y, Shomura T, Ogawa Y, Tsuruoka T, Watanabe H, Totsukawa K, Suzuki T, Moriyama C, Yoshida J, Inouye S, Niida T (1973) Sci Rep of Meiji Seika Kaisha 13:34 b) Omura S, Hinotozawa K, Imamura N, Murata M (1984) J Antibiot 27:939
8. a) Wohlleben W, Arnold W, Behrmann I, Broer I, Hillemann D, Pühler A, Strauch E (1991) Genetic analysis of different resistant mechanisms against the herbicidal antibiotic

phosphinothricyl-alanyl-alanin. In Baumberg S (ed) Genetics and product formation in streptomyces. Plenum Press, New York, p 171 b) Murakami T, Anzai H, Imai S, Satoh A, Nagaska K, Thompson CJ (1986) Mol Gen Genet 205:42 c) Thompson CJ, Morra NR, Tigard R, Crameri R, Davies JE, Lauwereys M, Botterman J (1987) EMBO Journal 6:2519
9. a) Botterman J, Leemans J (1989) Brit Crop Prot Conf Monogr 42:63; b) Wohlleben W, Arnold W, Broer I, Hillemann D, Strauch E, Pühler A (1988) Gene 70:25
10. Burg RW, Miller BM, Baker EE, Birbaum J, Currie SA, Harman R, Kong Y-L, Monaghan RL, Olsen G, Putter I, Tunae JB, Wallick H, Stapley EO, Oiura R, Omura S (1979) Antimicrob Agents and Chemother 15:361
11. Baker R, Swain J (1989) Chem Britain 25:692
12. a) Anke T, Oberwinkler F, Steglich W, Schramm G (1977) J Antibiot 30:806; b) Anke T, Steglich W (1989) ß-Methoxyacrylate antibiotics: from biological activity to synthetic analogues. In Schlunegger UP (ed) Biologically active molecules - identification, characterization, and synthesis. Springer Verlag, Berlin and Heidelberg, p 90 c) Clough JM (1993) Nat Prod Rep 10:565
13. Sauter, Ammermann, Roehl (1995) In Copping (ed) Crop protection agents from nature: Natural products and analogues. Royal Soc Chemistry, Cambridge
14. Petsko GA (1996) Nature 384 (Supp):7
15. Vandamme EJ (ed) (1984) Biotechnology of Industrial Antibiotics. Drugs and Pharmaceutical Sciences, vol 22. Marcel Dekker, New York
16. Goodfellow M, Williams ST, Mordarski (eds) (1988) Actinomycetes in Biotechnology. Academic Press, London
17. Omura S (ed) (1992)The Search for Bioactive Compounds from Microorganisms. Brock/Springer, New York
18. Gräfe U (1992) Biochemie der Antibiotika. Spektrum Acad Verl, Heidelberg
19. Yarbrough GG, Taylor DP, Rowlands RT, Crawford MS, Lasure LL (1993) J Antibiot 46:535
20. Grabley S, Thiericke R, Zeeck A (1994) Antibiotika und andere mikrobielle Wirkstoffe. In: Präve P, Faust U, Sittig W, Sukatsch DA (eds) Handbuch der Biotechnologie, 4th ed Oldenbourgverlag, München, p 663
21. Kuhn W, Fiedler H-P (eds) (1995) Sekundärmetabolismus bei Mikroorganismen, Beiträge zur Forschung (engl). Attempto, Tübingen
22. Frydrych CH (1993) Amino Acids Pep 24:245
23. a) Tillotson GS (1996) J Med Microbiol 44:320; b) Fass RJ, Barnishan J, Solomon MC, Ayers LW (1996) Antimicrob Agents Chemother 40:1412
24. Dreyfuss M, Harri E, Hofmann H (1976) Eur J Appl Micobiol 3:125
25. Borel FF, Feurer C, Gubler HV, Stähelin H (1976) Agents Actions 6:468
26. Borel FF, Feurer C, Magnée C, Stähelin H (1979) Immunology 32:1017
27. a) Rehácek Z (1995) Folia Microbiol 40:68; b) Stähelin HF (1996) Experientia 52:5
28. Kino T, Hatanaka H, Hashimoto M, Nishiyama M, Goto T, Okuhara M, Kohsaka, Aoki H, Imanaka HAT (1987) J Antibiot 40:1249
29. Tanaka H, Kuroda A, Marusawa H, Hatanaka H, Kino T, Goto T, Hashimoto M, Taga T (1987) J Am Chem Soc 109:5031
30. Baker H, Sidorowicz A, Sehgal SN, Vezina C (1978) J Antibiot 31:539
31. Fischer G (1994) Angew Chem 106:1479
32. Alberts AW, Chen J, Kuron G, Hunt V, Huff J, Hoffman C, Rothrock J, Lopez M, Joshua H, Harris E, Patchett A, Monaghan R, Currie S, Stapley E, Albers-Schonberg G, Hensens O, Hirschfield J, Hoogsteen K, Liesch J, Springer J (1980) Proc Natl Acad Sci USA 77:3957
33. Grundy SM (1978) West J Med 128:13
34. Endo A, Kuroda M, Tsujita Y (1976) J Antibiot 29:1346
35. Endo A, Tsujita Y, Kuroda M, Tanzawa K (1977) Eur J Biochem 77:31
36. Brown AG, Smale TC, King TJ, Hasenkamp R, Thompson RH (1976) J Chem Soc Perkin Transl, 1165
37. Endo A (1979) J Antibiot 32:852
38. Endo A (1980) J Antibiot 33:334
39. Endo A (1985) J Med Chem 28:401

40. Tanzawa K, Endo A (1979) Eur J Biochem 98:195
41. Nakamura CE, Abeles EH (1985) Biochem 24:1364
42. Stokker GE, Hoffman WF, Alberts AW, Cragoe EJ, Deana AA, Gilfillan JL, Huff JW, Novello FC, Prugh JD, Smith RL, Willard AK (1985) J Med Chem 28:347
43. Hoffman WF, Alberts AW, Cragoe EJ, Deana AA, Evans BE, Gilfillan JL, Gould NP, Huff JW, Novello FC, Prugh JD, Rittle KE, Smith RL, Stokker GE, Willard AK (1986) J Med Chem 29:159
44. Stokker GE, Alberts AW, Anderson PS, Cragoe EJ, Deana AA, Gilfillan JL, Hirschfield J, Holtz WJ, Hoffman WF, Huff JW, Lee TJ, Novello FC, Prugh JD, Rooney CS, Smith RL, Willard AK (1986) J Med Chem 29:170
45. Stokker GE, Alberts AW, Gilfillan JL, Huff JW, Smith RL (1986) J Med Chem 29:852
46. Bartmann W, Beck G, Granzer E, Jendralla H, Kerekjarto BV, Wess G (1986) Tetrahedron Lett 27:849
47. Stokker GE, Rooney CS, Wiggins JM, Hirschfield J (1986) J Org Chem 51:4931
48. Heathcock CH, Hadley CR, Rosen T, Theisen PD, Hecker SJ (1987) J Med Chem 30:1858
49. a) Yamamoto A, Itoh S, Hoshi K, Ichihara K (1995) Experientia 51:223 b) Plosker GL, Wagstaff AJ (1996) Drugs 51:433
50. Thomas G, Paglia R (1992) XI Int Symp Drugs Affecting Lipid Metabolism. Abstracts, p 69
51. Frommer W, Junge B, Müller L, Schmidt D, Truscheit E (1979) Planta Med 35:195
52. Truscheit E, Hillebrand J, Junge B, Müller L, Puls W, Schmidt D (1988) Prog Clin Biochem Med 7:17
53. Bischoff H (1994) Europ J Clin Invest 24 (Suppl 3):11
54. Bischoff H, Ahr HJ, Schmidt D, Stoltefuß J (1994) Nachr Chem Tech Lab 42:1119
55. Ahr HJ, Boberg M, Brendel E, Krause HP, Steinke W (1997) Arzneim-Forsch 47:734
56. Tormo MA, Ropero MF, Nieto M, Martinez IM, Campillo JE (1996) Can J Physiol Pharmacol 74:1196
57. Wani MC, Taylor HL, Wall ME, Coggin P, McPhail AT (1971) J Am Chem Soc 93:2325
58. Schiff PB, Fant J, Horwitz SB (1979) Nature (London) 277:665
59. Parnass J, Kingston DGI, Powell RG, Harracksingh C, Horwitz SB (1982) Biochm Biophys Res Commun 105:182
60. Wall ME (1993) In: Chronicles of Drug Discovery, Vol 3 (ed. Lednicer D) ACS Professional Reference Book, Washington, DC, p 327
61. Yukimune Y, Tabata H, Higashi Y, Hara Y (1996) Nature Biotechnol 14:1129
62. Stierle A (1993) Science 260:214
63. Nicolaou KC, Dai W-M, Guy RK (1994) Angew Chem 106:38
64. Nicolaou KC, Guy RK (1995) Angew Chem 107:2247
65. Tang W, Eisenbrand G (1992) Chinese Drugs of Plant Origin. Springer Verlag, Berlin, p 239
66. Wall ME, Wani MC, Cook CE, Palmer K, McPhail AT, Sim GA (1966) J Am Chem Soc 88:3888
67. Giovanella BC, Stehlin JS, Wall ME, Wani MC, Nicholas AW, Liu LF, Silber R, Potmesil M (1989) Science 246:1046
68. Slichenmyer WJ, Rowinsky EK, Donehower RC, Kaufmann SH (1993) J Natl Cancer Inst 85:271
69. Wall ME, Wani MC (1993) ACS Symp Ser (Human Medicinal Agents from Plants) 534:149
70. Klayman DL (1993) ACS Symp Ser (Human Medicinal Agents from Plants) 534:242
71. Schmid G, Hofheinz W (1983) J Am Chem Soc 105:624
72. Klayman DL (1985) Science 228:1049
73. Hien T, White NJ (1993) Lancet 341:603
74. Luo XD, Shen CC (1987) Med Res Rev 7:29
75. Ishida N, Miyazaki K, Kumagai K, Rikimaru M (1965) J Antibiot 18:68
76. Konishi M, Ohkuma H, Saitoh K, Kawaguchi H, Golik J, Dubay G, Gronewold G, Krishnan B, Doyle TW (1985) J Antibiot 38:1605

77. Lee MD, Dunne TS, Siegel MM, Chang CC, Morton GO, Borders DB (1987) J Am Chem Soc 109:3464
78. Konishi M, Ohkuma H, Matsumoto K, Tsuno T, Kamei H, Miyaki T, Oki T, Kawaguchi H, Van Duyne GD, Clardy J (1989) J Antibiot 42:1449
79. Silverman RB (1992) The Organic Chemistry of Drug Design and Drug Action, Academic Press, San Diego, USA, p 262
80. Smith AL, Nicolaou KC (1996) J Med Chem 39:2103
81. Martin DG, Biles C, Gerpheide SA, Hanka LJ, Krueger WC, McGovern JP, Mizsak SA, Neil GL, Stewart JC, Visser J (1981) J Antibiot 34:1119
82. Chidester CG, Krueger WC, Mizsak SA, Duchamp DJ, Martin DG (1981) J Am Chem Soc 103:7629
83. Scahill TA, Jensen RM, Swenson DH, Hatzenbuhler NT, Petzold G, Wierenga W, Brahme ND (1990) Biochem 29:2852
84. Lin CH, Beale JM, Hurley LH (1991) Biochem 30:3597
85. Sun D, Hurley LH (1992) J Med Chem 35:1773
86. Sun D, Hurley LH (1993) J Am Chem Soc 115:5925
87. Kelly RC, Aristoff PA (Upjohn Co.) US 88 243350 880912
88. Lee CS, Gibson NW (1991) Cancer Res 51:6586
89. Sun D, Hurley LH (1993) J Am Chem Soc 115:5925
90. Seaman FC, Hurley L (1996) J Am Chem Soc 118:10052
91. Woynarowski JM, Beerman TA (1997) Biochim Biophys Acta 1353:50
92. Carter CA, Waud WR, Li LH, DeKoning TF, Mc Govren JP, Plowman J (1996) Clin Cancer Res 2:1143
93. Bollag DM, McQuency PA, Zhu J, Hensens O, Kouplal L, Liesch J, Goetz M, Lazarides E, Woods CM (1995) Cancer Res 55:2325
94. Gerth K, Bedorf N, Höfle G, Reichenbach H (1996) J Antibiot 49:560
95. Höfle G, Bedorf N, Steinmetz H, Schomburg D, Gerth K, Reichenbach H (1996) Angew Chem 35:1567
96. a) Balog A, Meng D, Kamenecka T, Bertinato P, Su DS, Sorensen EJ, Danishefsky SJ (1996) Angew Chem 108:23 b) Yang Z, He Y, Vourloumis D, Vallberg H, Nicolaou KC (1997) Angew Chem 109:170 c) Nicolaou KG, Sarabia F, Ninkovic S, Yang Z (1997) Angew Chem 109:543
97. Schinzer D, Limberg A, Bauer A, Böhm OM, Cordes M (1997) Angew Chem 109:543
98. Su DS, Meng D, Bertinato P, Balog A, Sorensen EJ, Danishefsky J, Zheng YH, Chou TC, He L, Horwitz SB (1997) Angew Chem 109:775
99. Bergstrom JD, Kurtz MM, Rew DJ, Amend AM, Karkas JD, Bostedor RG, Bansal VS, Dufresne C, Van-Middlesworth FL, Hensens OD, Liesch JM, Zink DL, Wilson KE, Onishi J, Milligan JA, Bills G, Kaplan L, Nallin-Omstead M, Jenkins RG, Huang L, Meinz MS, Quinn L, Burg RW, Kong YL, Mochales S, Mojena M, Martin I, Pelaez F, Diez MT, Alberts AW (1993) Proc Natl Acad Sci USA 90:80
100. a) Dawson MJ, Farthing JE, Marshall PS, Middleton RF, O'Neill MJ, Shuttleworth A, Stylli C, Tait RM, Taylor PM, Wildman HG, Buss AD, Langley D, Hayes MV (1992) J Antibiot 45:639; b) Sidebottom PJ, Highcock RM, Lane SJ, Procopiou PA, Watson NS (1992) J Antibiot 45:648
101. Hasumi K, Tachikawa K, Sakai K, Murakawa S, Yoshikawa N, Kumazawa S, Endo A (1993) J Antibiot 46:689
102. Nadin A, Nicolaou KC (1996) Angew Chem 108:1732
103. a) Tang XG, De Sarno R, Sugaya K, Giacobini E (1989) J Neurosci Res 24:276; b) Borman S (1993) Chem & Eng News, September 20, p 35
104. Liu JS, Zhou YZ, Yu CM, Zhou YZ, Han YY, Wu FW, Qi BF (1986) Can J Chem 64:837
105. Kozikowski AP, Xia Y, Reddy ER, Tückmantel W, Hanin I, Tang XC (1991) J Org Chem 56:4636
106. Attaway DH, Zaborsky OR (1993) Marine Biotechnology, Vol 1, Pharmaceutical and Bioactive Natural Products. Plenum Press, New York
107. a) de Vries DJ, Beart PM (1995) TiBS 16:275; b) König GM, Wright AD (1996) Planta Med 62:193

108. Jensen PR, Fenical W (1996) J Ind Microbiol 17:346
109. Yasumoto T, Bagnis R, Venoux JP (1976) Bull Jpn Soc Sci Fish, 42:359
110. a) Sasaki M, Matsumori N, Maruyama T, Nonomura T, Murata M, Tachibana K, Yasumoto T (1996) Angew Chem 108:1782; b) Nonomura T, Sasaki M, Matsumori N, Murata M, Tachibana K, Yasumoto T (1996) Angew Chem 108:1786
111. Satake M, Ishida S, Yasumoto T, Murata M, Utsumi H, Hinomoto T (1995) J Am Chem Soc 117:7019
112. Murata M, Naoki H, Iwashita T, Matsunaga S, Sasaki M, Yokoyama A, Yasumoto T (1993) J Am Chem Soc 115:2060
113. Gusovsky F, Daly JW (1990) Biochem Pharmacol 39:1633
114. Lin YY, Risk M, Ray SM, Van Engen D, Clardy J, Golik J, James JC, Nakanishi K (1981) J Am Chem Soc 103:6773
115. Lombet A, Bidard J-N, Lazdunski M (1987) FEBS Lett 219:355
116. Kao CY, Levinson SR (eds) Tetrodotoxin, saxitoxin, and the molecular biology of the sodium channel (1986) The New York Academy of Sciences, New York
117. Catterall WA, Risk M (1981) Mol Pharmacol 19:345
118. Yasumoto T, Murata M (1993) Chem Rev 93:1897
119. Tachibana K, Scheuer PJ, Tsukitani Y, Kikuchi H, Van Engen D, Clardy J, Gopichand Y, Schmitz FJ (1981) J Am Chem Soc 103:2469
120. Murakami Y, Oshima Y, Yasumoto T (1982) Nippon Suisan Gakkaishi 48:69
121. Haystead TAJ, Sim ATR, Carling D, Honnor RC, Tsukitani Y, Cohen P, Hardie DG (1989) Nature 337:78
122. Moore RE (1985) Prog Chem Org Nat Prod 48:81
123. Li W-R, Ewing WR, Harris BD, Joullié MM (1990) J Am Chem Soc 112:7659
124. Crews CM, Collins JL, Lane WS, Snapper ML, Schreiber SL (1994) J Biol Chem 269:15411
125. Sakai R, Rinehart KL, Kishore V, Kundu B, Faircloth G, Gloer JB, Carney JR, Namikoshi M, Sun F, Hughes RG, Grávalos DG, de Quesada TG, Wilson GR, Heid RM (1996) J Med Chem 39:2819
126. Pettit GR, Herald CL, Doubek DL, Herald DL (1982) J Am ChemSoc 104:6846
127. Berkow RL, Kraft AS (1985) Biophys Res Commun 131:1109
128. Hennings H, Blumberg PM, Pettit GR, Herald CL, Shores R, Yuspa SH (1987) Carcinogenesis 8:1343
129. Schaufelberger DE, Koleck MP, Beutler JA, Vatakis AM, Alvarado AB, Andrews P, Marzo LV, Muschik GM, Roach J, Ross JT, Lebherz WB, Reeves MP, Eberwein RM, Rodgers LL, Testerman RP, Snader KM, Forenza S (1991) J Nat Prod 54:1265
130. Rinehart KL, Holt TG, Fregeau NL, Stroh JG, Keifer PA, Sun F, Li LH, Martin DG (1990) J Org Chem 55:4512
131. Moore BM, Seaman FC, Hurley LH (1997) J Am Chem Soc 119:5475
132. Glaser KB, de Carvahlo MS, Jacobs RS, Kernan MR, Faulkner DJ (1989) Mol Pharmacol 36:782
133. Potts BCM, Faulkner DJ, de Carvahlo MS, Jacobs RS (1992) J Am ChemSoc 114:5093
134. Look SA, Fenical W, Matsumoto GK, Clardy J (1986) J Org Chem 51:5140
135. Roussis V, Wu Z, Fenical W, Strobel SA, Van Duyne GD, Clardy J (1990) J Org Chem 55:4916
136. Schultz M, Tsaklakidis C (1997) Nach Chem Tech Lab 45:159
137. Omura S (1992) The search for bioactive compounds from microorganisms. Omura S (ed), Springer, New York
138. Umezawa S, Tsuchiya T, Tatsuta K, Horiuchi Y, Usui T (1970) J Antibiot 23:20
139. Umezawa S, Usui T, Umezawa H, Tsuchiya T, Takeuchi T, Hamada M (1971) J Antibiot 24:85
140. Bindseil KU, Henkel T, Zeeck A, Bur D, Niederer D, Séquin U (1991) Helv Chim Acta 74:1281
141. Drautz H, Zähner H (1981) Zbl Bakt Suppl 11:515
142. Zähner H (1978) In Hütter R, Leisinger T, Nüesch J, Wehrli W (eds) Antibiotics and Other Secondary Metabolites. Biosynthesis and Production. Academic Press, London, New York, San Francisco, p 1

143. Grabley S, Wink J, Zeeck A (1990) In: Biotechnology Focus 3. Finn RK, Präve P, Schlingmann M, Crueger W, Esser K, Thauer R, Wagner F (eds), Hanser Publ. Munich, Vienna, New York, Barcelona, p 359
144. Nakagawa A (1992) In: The search for bioactive compounds from microorganisms. Omura S (ed), Springer, New York, p 263
145. Noltemeyer M, Sheldrick GM, Hoppe H-U, Zeeck A (1982) J Antibiot 35:549
146. Chem Y, Zeeck A, Chen Z, Zähner H (1983) J Antibiot 36:913
147. Fiedler H-P (1984) J Chromatogr 316:487
148. Fiedler H-P, Kulik A, Schütz T, Volkmann C, Zeeck A (1994) J Antibiot 47:1116
149. Volkmann C, Hartjen U, Zeeck A, Fiedler HP (1995) J Antibiot 48:522
150. Pfefferle C, Kempter C, Metzger J, Fiedler HP (1996) J Antibiot 49:826
151. Morino T, Nishimoto M, Masuda A, Fujita S, Nishikiori T, Saito S (1995) J Antibiot 48:1509
152. Blum S, Fiedler HP, Groth I, Kempter C, Stephan H, Nicholson G, Metzger J, Jung, G (1995) J Antibiot 48:619
153. Rodriguez S, Wolfender J-L, Hostettmann K, Odontuya G, Purev O (1996) Helv Chim Acta 79:363
154. Gu Z-M, Zhou D, Wu J, Shi G, Zeng L, McLaughlin JL (1997) J Nat Prod 60:242
155. Hostettmann K, Potterat O, Wolfender J-L (1997) Pharm Ind 59:339
156. Fiedler HP, Nega M, Pfefferle C, Groth I, Kempter C, Stephan H, Metzger J (1996) J Antibiot 49:758
157. Stephan H, Kempter C, Metzger J, Jung, G, Potterat O, Pfefferle C, Fiedler H-P (1996) J Antibiot 49:765
158. Burkhardt K, Fiedler HP, Grabley S, Thiericke R, Zeeck A (1996) J Antibiot 49:432
159. Schneider A, Späth J, Breding-Mack S, Zeeck A, Grabley S, Thiericke R (1996) J Antibiot 49:438
160. Göhrt A, Grabley S, Thiericke R, Zeeck A (1996) Liebigs Ann Chem 627
161. Grabley S, Thiericke R, Zerlin M, Göhrt A, Philipps S, Zeeck A (1996) J Antibiot 49:593
162. Fuchser J, Grabley S, Noltemeyer M, Philipps S, Thiericke R, Zeeck A (1994) Liebigs Ann Chem 831
163. Grote R, Zeeck A, Drautz H, Zähner H (1988) J Antibiot 41:1178
164. Grote R, Zeeck A, Beale Jr. JM (1988) J Antibiot 41:1186
165. Grabley S, Granzer E, Hütter K, Ludwig D, Mayer M, Thiericke R, Till G, Wink J (1992) J Antibiot 45:56
166. Göhrt A, Zeeck A, Hütter K, Kirsch R, Kluge H, Thiericke R (1992) J Antibiot 45:66
167. Grabley S, Hammann P, Hütter K, Kirsch R, Kluge H, Thiericke R, Mayer M, Zeeck A (1992) J Antibiot 45:1176
168. Mayer M, Thiericke R (1993) J Antibiot 46:1372
169. Dräger G, Kirschning A, Thiericke R, Zerlin M (1996) Nat Prod Rep 13:365
170. Rohr J, Thiericke R (1992) Nat Prod Rep 9:103
171. Drautz H, Reuschenbach P, Zähner H, Rohr J, Zeeck A (1985) J Antibiot 38:1292
172. Fiedler H-P, Rohr J, Zeeck A (1986) J Antibiot 39:856
173. Bach G, Breiding-Mack S, Grabley S, Hammann P, Hütter K, Thiericke R, Uhr H, Wink J, Zeeck A (1993) Liebigs Ann Chem 241
174. Grabley S, Thiericke R, Wink J, Henne P, Philipps S, Wessels P, Zeeck A (1994) J Nat Prod 57:541
175. Schönewolf M, Grabley S, Hütter K, Machinek R, Wink J, Zeeck A, Rohr J (1991) Liebigs Ann Chem 77
176. Schönewolf M, Rohr J (1991) Angew Chem 103:211
177. Henkel T, Zeeck A (1991) Liebigs Ann Chem 367
178. Tanaka Y, Kanaya I, Takahashi Y, Shinose M, Tanaka H, Omura S (1993) J Antibiot 46:1208
179. Tanaka Y, Kanaya I, Shiomi K, M, Tanaka H, Omura S (1993) J Antibiot 46:1214
180. Omura S, Tanaka Y, Kanaya I, Shinose M, Takahashi Y (1990) J Antibiot 43:1034
181. Henkel T, Breiding-Mack S, Zeeck A, Grabley S, Hammann PE, Hütter K, Till G, Thiericke R, Wink J (1991) Liebigs Ann Chem 575

182. Grabley S, Kretzschmar G, Mayer M, Philipps S, Thiericke R, Wink J, Zeeck A (1993) Liebigs Ann Chem 573
183. Mayer M, Thiericke R (1993) J Chem Soc Perkin Trans 1 2525
184. Zerlin M, Thiericke R (1994) J Org Chem 59:6986
185. Hoff H, Drautz H, Fiedler H-P, Zähner H, Schultz JE, Keller-Schierlein W, Philipps S, Ritzau M, Zeeck A (1992) J Antibiot 45:1096
186. Ritzau M, Philipps S, Zeeck A, Hoff H, Zähner H (1993) J Antibiot 46:1625
187. Thiericke R, Zerlin M (1996) Nat Prod Lett 8:163
188. Ritzau M, Keller M, Wessels P, Stetter KO, Zeeck A (1993) Liebigs Ann Chem 871
189. Henne P, Grabley S, Thiericke R, Zeeck A (1997) Liebigs Ann/Recueil 937
190. Drautz H, Zähner H, Rohr J, Zeeck A (1986) J Antibiot 39:1657
191. Rohr J, Zeeck A (1987) J Antibiot 40:459
192. Henne P, Thiericke R, Grabley S, Hütter K, Wink J, Jurkiewicz E, Zeeck A (1993) Liebigs Ann Chem 565
193. Vreven F (1997) Labo, July, p 48
194. Koch C, Neumann T, Thiericke R, Grabley S (1997) Nachr Chem Tech Lab 45:16
195. Koch C, Neumann T, Thiericke R, Grabley S (1997) BIOspektrum 3:43
196. Nefzi A, Ostresh JM, Houghten RA (1997) Chem Rev 97:449
197. Balkenhohl F, von dem Bussche-Hünnefeld C, Lansky A, Zechel C (1996) Angew Chem 108:2436
198. Früchtel JS, Jung G (1996) Angew Chem 108:19
199. Gordon EM, Gallop MA, Patel DV (1996) Acc Chem Res 29:144
200. Ecker DJ, Crooke ST (1995) Biotechnol 13:351
201. Terrett NK, Gardner M, Gordon DW, Kobylecki, Steele J (1995) Tetrahedron 51:8135
202. Gallop MA, Barrett W, Dower WJ, Fodor SPA, Gordon EM (1994) J Med Chem 37:1232
203. Gordon EM, Darrett RW, Dower WJ, Fodor PA, Gallop MA (1994) J Med Chem 37:1385
204. Hermkens PHH, Ottenheijm HCJ, Rees D (1997) Tetrahedron 52:4527
205. Hermkens PHH, Ottenheijm HCJ, Rees D (1997) Tetrahedron 53:5643
206. Thompson LA, Ellman JA (1996) Chem Rev 96:555
207. Yun W, Mohan R (1996) Tetrahedron Lett 37:7189
208. Arumugam V, Routledge A, Abell C, Balasubramanian S (1997) Tetrahedron Lett 38:6473
209. Collini MD, Ellingboe JW (1997) Tetrahedron Lett 38:7963
210. Hutchins SM, Chapman KT (1996) Tetrahedron Lett 37:4869
211. Kim RM, Manna M, Hutchins SM, Griffin PR, Yates NA, Bernick AM, Chapman KT (1996) Proc Natl Acad Sci USA 93:10012
212. Hughes I (1996) Tetrahedron Lett 37:7595
213. Cheng Y, Chapman KT (1997) Tetrahedron Lett 37:1497
214. Kaljuste K, Undén A (1995) Tetrahedron Lett 36:9211
215. Mayer JP, Bankaitis-Davis D, Zhang J, Beaton G, Bjergarde K, Andersen CM, Goodman BA, Herrera CJ (1996) Tetrahedron Lett 37:5633
216. Mohan R, Chu Y-L, Morrissey MM (1996) Tetrahedron Lett 37:3963
217. Yang L, Guo L (1996) Tetrahedron Lett 37:5041
218. Gopalsamy A, Pallai PV (1997) Tetrahedron Lett 38:907
219. Ruhland T, Künzer H (1996) Tetrahedron Lett 37:2757
220. Kiselyov AS, Amstrong RW (1997) Tedrahedron Lett 38:6163
221. Meutermans WDF, Alewood PF (1995) Tetrahedron Lett 36:7709
222. Rölfing K, Thiel M, Künzer H (1996) Synlett 1036
223. Hutchins SM, Chapman KT (1996) Tetrahedron Lett 37:4865
224. Nicolaou KC, Winssinger N, Pastor J, Ninkovic S, Sarabia F, He Y, Vourloumis D, Yang Z, Li T, Giannakakou P, Hamel E (1997) Nature 387:268
225. Nicolaou KC, Vourloumis D, Li T, Pastor J, Winssinger N, He Y, Ninkovic S, Sarabia F, Vallberg H, Roschangar F, King NP, Finlay MRV, Giannakakou P, Verdier-Pinard DP, Hamel E (1997) Angew Chem 109:2181
226. Nicolaou KC, Sarabia F, Finlay MRV, Ninkovic S, King NP, Vourloumis D, He Y (1997) Chem Eur J 3:1971

227. Nielsen J, Lyngsf LO (1996) Tetrahedron Lett 37:8439
228. IBC Conference on Combinatorial Synthesis of Patural Products, Dec. 1997, San Francisco, USA
229. Chen S, Janda KD (1997) J Am Chem Soc 119:8724
230. Reggelin M, Brenig V, Welcker R, Tetrahedron Lett, submitted
231. Keinan E, Sinha A, Yazbak A, Sinha Santosh C, Sinha Subhash C (1997) Pure Appl Chem 69:423
232. Rohr J (1995) Angew Chem Int Ed Eng 34:881
233. Kasal A, Kohout L, Lebl M (1995) Collect Czech Chem Commun 60:2147
234. Hirschmann R, Sprengeler PA, Kawasaki T, Leathy EM, Shakespeare WC, Smith III AB (1993) Tetrahedron 49:3665
235. Boyce R, Li G, Nestler P, Suenaga T, Still WC (1994) J Am Chem Soc 116:7955
236. Wess G, Bock K, Kleine H, Kurz M, Guba W, Hemmerle H, Lopez-Calle E, Baringhaus K-H, Glombik H, Ehnsen A, Kramer W (1996) Angew Chem Int Ed Engl 35:2222
237. Xiao XY (1997) IBC Conference on Combinatorial Synthesis of Patural Products, Dec. 1997, San Francisco, USA
238. Atuegbu A, Maclean D, Nguyen C, Gordon EM, Jacobs JW (1996) Bioorg Med Chem 4:1097
239. Daum SJ, Lemke JR (1979) Annu Rev Microbiol 33:241
240. Hutchinson CR (1988) Med Res Rev 8:5571
241. Savidge TA (1984) In Vandamme EJ (ed) Biotechnology of Industrial Antibiotics. Marcel Dekker, New York, p 171
242. Mahato SB, Garai S (1997) Steroids 62:332
243. Murray HC, Peterson DH (1952) US Patent 2602769 (Upjohn Co, Kalamazoo, Michigan, USA)
244. Kieslich K (1976) In Microbial Transformations of Non-Steroid Compounds. Thieme, Stuttgart
245. Rosazza JP (1982) In Microbial Transformations of Bioactive Compounds. CRC Press, Boca Raton
246. Kieslich K (ed) Biotransformations, Vol 6a of Rehm HJ, Reed G (eds) Biotechnology, A Comprehensive Treatise in 8 Volumes. Verlag Chemie, Weinheim
247. Yamada H, Shimizu S (1988) Angew Chem 100:640
248. Sih C, Chen C (1984) Angew Chem 96:556
249. Andersson E, Halm-Hägerdal B (1990) Enzyme Micro Technol 242
250. Hardmann DJ (1991) Crit Rev Biotechnol 11:1
251. Kieslich K (1991) Acta Biotechnol 11:559
252. Davies HG, Green RH, Kelly DR, Roberts SM (1989) In: Biotransformations in Preparative Organic Chemistry. Academic Press, London
253. Sariaslani FS, Rosazza JPN (1984) Enzyme Microb Technol 6:242
254. Aparicio JF, Molnar I, Schwecke T, König A, Haydock SF, Khaw LE, Staunton J, Leadlay PF (1996) Gene 169:9
255. Patel RN (1997) Adv Appl Microbiol 43:91
256. Turner NJ (1989) Nat Prod Rep 6:626
257. Kuhnt M, Bitsch F, France J, Hofman H, Sanglier JJ, Traber R (1996) J Antibiot 49:781
258. Schulman M, Doherty P, Zink D, Arison B (1993) J Antibiot 46:1016
259. Nishida H, Sakakibara T, Aoki F, Saito T, Ichikawa K, Inagaki T, Kojima Y, Yamauchi Y, Huang LH, Guadliana MA, Kaneko T, Kojima N (1995) J Antibiot 48:657
260. Mouslim J, Cuer A, David L, Tabet JC (1995) J Antibiot 48:1011
261. Ajito K, Kurihara KI, Shibahara S, Hara O, Shimizu A, Araake M, Omoto S (1997) J Antibiot 50:92
262. Thiericke R, Rohr J (1993) Nat Prod Rep 10:265
263. Ogata K, Osawa H, Tani Y (1977) J Ferment Technol 55:285
264. Häggblom P, Niehaus WG (1987) Exp Mycology 11:150
265. Thiericke R, Zerlin M (1996) Nat Prod Lett 8:163
266. Zerlin M, Thiericke R, Henne P, Zeeck A (1997) Nat Prod Lett 10:217

267. Liefke E, Onken U (1992) Biotechnol Bioeng 40:719
268. Onken U, Liefke E (1989) Adv Biochem Eng Biotechnol 40:137
269. Liefke E, Kaiser D, Onken U (1990) Appl Microbiol Biotechnol 32:674
270. Zeeck A, Sattler I, Boddien C (1993) DECHEMA Monogr 129:85
271. Kaiser D, Onken U, Sattler I, Zeeck A (1994) Appl Microbiol Biotechnol 41:309
272. Dick O, Onken U, Sattler I, Zeeck A (1994) Appl Microbiol Biotechnol 41:373
273. Arison BH, Omura S (1974) J Antibiot 27:28
274. DAgnolo G, Rosenfeld IS, Awaya J, Omura S, Vagelos PR (1973) Biochim Biophys Acta 326:155
275. Kawaguchi A, Tomoda H, Nozoe S, Omura S, Okuda S (1982) J Biochem 92:7
276. Funabashi H, Kawaguchi A, Tomada H, Omura S, Okuda S, Iwasaki S (1989) J Biochem 105:751
277. Omura S (1976) Bacteriol Rev 40:681
278. Omura S, Nakaga A, Taheshima H, Atsumi K, Miyazawa J, Pirou F, Lukacs G (1975) J Am Chem Soc 97:6600
279. Omura S (1984) Macrolide antibiotics Chemistry, Biology, and Practice, Academic Press, New York
280. Nakagawa A, Omura S (1995) J Antibiot 49:717
281. Omura S, Sadakane N, Tanaka Y, Matsubara H (1983) J Antibiot 36:927
282. Sadakane N, Tanaka Y, Omura S (1983) J Antibiot 36:921
283. Demain AL, Ninet L, Bost PE, Bouanchaud DH, Florent J (1981) In: The Future of Antibiotherapy and Antibiotic Research. Academic Press, London, p 417
284. Birch AJ (1963) Pure Appl Chem 7:527
285. Rinehart KL Jr (1977) Pure Appl Chem 49:1361
286. Queener SW, Sebek OK, Vézina C (1978) Ann Rev Microbiol 32:593
287. Spagnoli R, Cappelletti L, Toscana L (1983) J Antibiot 36:356
288. Omura S, Kitao C, Matsubara H (1980) Chem Pharm Bull 28:1963
289. Johdo O, Yoshimoto A, Ishikura T, Naganawa H, Takeuchi T, Umezawa H (1986) Agric Biol Chem 50:1657
290. Nakagawa M, Hayakawa Y, Imamura K, Seto H, Otake N (1985) J Antibiot 38:821
291. Wagner C, Eckhardt K, Schuhmann G, Ihn W, Tresselt D (1984) J Antibiot 37:691
292. Hoshino T, Setoguchi Y, Fujiwara A (1984) J Antibiot 37:1469
293. Yoshimoto A, Johdo O, Takatsuke Y, Ishikura T, Sawa T, Takeuchi T, Umezawa H (1984) J Antibiot 37:935
294. Tobe H, Yoshimoto A, Ishikura T, Naganawa H, Takeuchi T, Umezawa H (1982) J Antibiot 35:1641
295. Oki T, Yoshimoto A, Matsuzawa T, Takeuchi T, Umezawa H (1980) J Antibiot 33:1331
296. Tynan III EJ, Nelson TH, Davies RA, Wernau WC (1992) J Antibiot 45:813
297. Kitamura S, Kase H, Odakura Y, Iida T, Shirahata K, Nakayama K (1982) J Antibiot 35:94
298. Leboul J, Davies J (1982) J Antibiot 35:527
299. Takeda K, Kinumaki A, Okuno S, Matsushita T, Ito Y (1978) J Antibiot 31:1039
300. Traxler P, Ghisalba O (1982) J Antibiot 35:1361
301. Cricchio R, Antonini P, Ferrari P, Ripamonti A, Tuan G, Martinelli E (1981) J Antibiot 34:1257
302. Lancini G, Hengeler C (1969) J Antibiot 22:637
303. Bormann C, Mattern C, Schrempf H, Fiedler HP, Zähner H (1989) J Antibiot 42:913
304. Bormann C, Huhn W, Zähner H, Rathmann R, Hahn H, König WA (1985) J Antibiot 38:9
305. Delzer J, Fiedler HP, Müller H, Zähner H, Rathmann R, Ernst K, König WA (1984) J Antibiot 37:80
306. Goeke K, Hoehn P, Ghisalba O (1995) J Antibiot 48:428
307. Ikeda H, Omura S (1995) J Antibiot 48:549
308. Gerlitz M, Udvarnoki G, Rohr J (1995) Angew Chem Int Ed Engl 34:1617
309. Rohr J, Schönewolf M, Udvarnoki G, Eckardt K, Schumann G, Wagner C, Beale JM, Sorey SD (1993) J Org Chem 58:2547
310. Rohr J (1995) Angew Chem Int Ed Engl 34:881

311. Khosla C, Zawada RJX (1996) TIBTech 14:335
312. Hopwood DA, Malpartida F, Kieser HM, Ikeda H, Duncan J, Fujii I, Rudd BAM, Floss HG, Omura S (1985) Nature 314:642
313. McDaniel R, Ebert-Khosla S, Hopwood DA, Khosla C (1993) J Am Chem Soc 115:11671
314. Scott AI (1992) Tetrahedron 48:2559
315. Floss HG, Strohl WR (1991) Tetrahedron 47:6045
316. Roberts SM, Turner NJ, Willetts AJ, Turner MK (1995) Introduction to Biocatalysis Using Enzymes and Micro-Organisms, Cambridge University Press
317. Dordick JS (1992) Biotech Progr 8:259
318. Elling L (1997) In: Scheper (ed) New Enzymes for Organic Synthesis. Springer Verlag, Berlin, p 145
319. Wong CH, Whitesides GM (1994) In: Baldwin JE, Magnus PD (eds) In Enzymes in Synthetic Organic Chemistry. Elsevier, New York, p 19
320. Khmelnitsky YL, Michels PC, Dordick JS, Clark DS (1996) In: Chaiken IM, Janda KD (eds) Molecular Diversity and Combinatorial Chemistry, Libraries and Drug Discovery. American Chemical Society, Washington DC, p144
321. Dordick JS (1997) In: High Throughput Screening for Drug Discovery, Cambridge Healthtech Institute Conference. Arlington, VA
322. Khmelnitsky YL, Budde C, Arnold JM, Usyatinsky A, Clark DS, Dordick JS (1997) J Am Chem Soc 119:11554
323. Ashton MJ, Jaye M, Mason JS (1996) Drug Discovery Today 1:11
324. Rommens JM, Iannuzzi MC, Kerem B, Drumm ML, Melmer G, Dean M, Rozmahel R, Cole JL, Kennedy D, Hidaka N (1989) Science, 245:1059; Riordan JR, Rommens JM, Kerem B, Alon N, Rozmahel R, Grzelczak Z, Zielenski J, Lok S, Plavsic N, Chou JL (1989) Science 245:1066
325. The Huntington's Disease Collaborative Research Group (1993) Cell 72:971
326. Zhang Y, Proenca R, Maffei M, Barone M, Leopold L, Friedman JM (1994) Nature 372:425
327. Adams MD, Kelley JM, Gocayne JD, Dubnick M, Polymeropoulos MH, Xiao H, Merril CR, Wu A, Olde B, Moreno RF, Kerlavage AR, McCombie WR, Venter JC (1991) Science 252:1651
328. Liang P, Pardee AB (1992) Science 257:967
329. Friedrich GA (1996) Nature Biotechnol 14:1234
330. Broach JR, Thorner J (1996) Nature 384 (Suppl):14
331. Kirsch DR (1993) Curr Opin Biotechnol 4:543
332. Palmer MA (1996) Nature Biotechnol 14:513
333. Hertzberg RP (1993) Curr Opin Biotechnol 4
334. Munder T, Ninkovic M, Rudakoff B (1997) Biotech Annu Rev 3:31
335. Osborne MA, Dalton S, Kochan JP (1995) Biotechnol 13:1474
336. Kralli A, Bohen SP, Yamamoto KR (1995) Proc Natl Acad Sci USA 92:4701
337. Picard D, Kursheed B, Garabedian MJ, Fortin MG, Lindquist S, Yamamoto KR (1990) Nature 348:166
338. Shiau SP, Glasebrook A, Hardiakar SD, Yang NN, Hershberger CL (1996) Gene 179:205
339. Coldham NG, Dave M, Sivapathasundaram S, McDonnell DP, Connor C, Sauer MJ (1997) Environmental Health Perspectives 105:734
340. Zerlin A (1997) Aufbau und Evaluierung von Testsystemen in *Saccharomyces cerevisiae*: Der humane Estrogen- und Androgenrezeptor und die Tax/CREB-interaktion als molekulare Targets. Ph-D Thesis, University of Jena, Germany
341. Mak P, McDonnell DP, Weigel NL, Schrader WT, O'Malley BW (1989) J Biol Chem 264:21613
342. Tran DQ, Klotz DM, Ladlie BL, Ide CF, McLachlan JA, Arnold SF (1996) Biochem Biophys Res Comm 229:518
343. Jin L, Tran DQ, Ide CF, McLachlan JA, Arnold SF (1997) Biochem Biophys Res Comm 233:139
344. McDonnell DP, Pike JW, Drutz DJ, Butt TR, O'Malley BW (1989) Mol Cell Biol 9:3517

345. Berghöfer-Hochheimer Y, Zurek, C, Langer G, and Munder T (1997) J Cell Biochem 65:184
346. Eigen M, Rigler R (1994) Proc Natl Acad Sci (USA) 91:5740
347. Hogan JC (1997) Nature Biotechnol 15:328
348. Bevan P, Ryder H, Shaw I (1995) TIBTECH 13:115
349. Hurst WJ ed. (1995) Automation in the Laboratory, VCH New York
350. Pharmaceutical Education & Research Institute, Project Management in the Pharmaceutical Industry, 1996
351. Centre for Medicines Research (CMS), Scrip´s Yearbook (several volumes), London
352. Major J (1996) Proceedings of the International Symposium on Laboratory Automation and Robotics, 25
353. Schmid I, Sattler I, Grabley S, Thiericke R (1997) Proceedings International Symposium on Laboratory Automation and Robotics.
354. Beals P (1995) In: Hurst WJ (ed), Automation in the Laboratory, VCH New York, p 109
355. Hurst WJ (1995) In: Hurst WJ (ed), Automation in the Laboratory, VCH New York, p 91
356. Burbaum JJ (1997) Proceedings of the International Symposium on Laboratory Automation and Robotics, in press
357. Elands J (1995) Proceedings of the International Symposium on Laboratory Automation and Robotics, 159
358. Green C (1996) Proceedings of the International Symposium on Laboratory Automation and Robotics, 164
359. Allee C (1996) Laboratory Robotics and Automation 8:307
360. Beese K (1996) Pharmaceutical bioprospecting and synthetic molecular diversity. Draft discussion paper, May 15

Protein Glycosylation: Implications for In Vivo Functions and Therapeutic Applications

Prakash K. Bhatia[1,3] · Asok Mukhopadhyay[2]

[1] National Institute of Immunology, Aruna Asaf Ali Marg, New Delhi-110 067, India
[2] National Institute of Immunology, Aruna Asaf Ali Marg, New Delhi-110 067, India.
E-mail: ashok@nii.ernet.in
[3] 499, Sector 11, Hiranmagri, Udaipur 313001, India

The glycosylation machinery in eukaryotic cells is available to all proteins that enter the secretory pathway. There is a growing interest in diseases caused by defective glycosylation, and in therapeutic glycoproteins produced through recombinant DNA technology route. The choice of a bioprocess for commercial production of recombinant glycoprotein is determined by a variety of factors, such as intrinsic biological properties of the protein being expressed and the purpose for which it is intended, and also the economic target. This review summarizes recent development and understanding related to synthesis of glycans, their functions, diseases, and various expression systems and characterization of glycans. The second section covers processing of *N*- and *O*-glycans and the factors that regulate protein glycosylation. The third section deals with in vivo functions of protein glycosylation, which includes protein folding and stability, receptor functioning, cell adhesion and signal transduction. Malfunctioning of glycosylation machinery and the resultant diseases are the subject of the fourth section. The next section covers the various expression systems exploited for the glycoproteins: it includes yeasts, mammalian cells, insect cells, plants and an amoeboid organism. Biopharmaceutical properties of therapeutic proteins are discussed in the sixth section. In vitro protein glycosylation and the characterization of glycan structures are the subject matters for the last two sections, respectively.

Keywords: Glycoprotein, Protein stability, Disease, Expression, Half-life.

Advances in Biochemical Engineering/
Biotechnology, Vol. 64
Managing Editor: Th. Scheper

Abbreviations

APP	Acute phase protein
Asn	Asparagine
ATP	Adenosine triphosphate
CD2	Clusters of diterminant
CG	Chorionic gonadotropin
CHO	Chinese hamster ovary

DolPP	Dolichol pyrophosphate
Endo F/H/D	Endopeptidase F/H/D
EPO	Erythropoietin
ER	Endoplasmic reticulum
FT	Fucosyl transferase
Fuc	Fucose
FSH	Follicular stimulating hormone
αG	α-Glucosidase
Glc	Glucose
Gal	Galactose
Gnc (GlcNAc)	*N*-acetylglucosamine
Glyc T	Glycosyl transferases
Gn T	*N*-acetylglucosaminyl transferase
Gal T	Galactosyl transferases
GalNAc	*N*-acetylgalactosamine
GalNAc T	*N*-acetylgalactosaminyl transferase
GPI	Glucosylphosphatidylinositol
IFNγ	Interferon γ
IgG	Immunoglobulin G
LH	Luteinizing hormone
Man	Mannose
α M	α-Mannosidase
PCR	Polymerase chain reaction
PNGases	Peptide *N*-glycanases
SA	Sialic acid
ST	Sialyl transferase
TSH	Thyroid stimulating hormone
tPA	Tissue plasminogen activator
UDP	Uridine diphosphate
Xyl	Xylose

1 Introduction

To obtain their final structural features and functions, newly formed polypeptides undergo various types of modifications, such as folding, conformation stabilization by disulfide bridges, assembly into homo- or hetero-oligomers, acquisition of prosthetic groups, specific proteolytic cleavage and covalent attachment of phosphate, sulfate, fatty acid, complex lipid or sugar groups [1]. The types of modification shown by an individual protein depend on its amino acid sequence, conformation, as well as on cell type and tissue context [2].

Glycoproteins are a conjugated form of proteins containing one or more heterosaccharides covalently bound to the polypeptide chain. These are present in virtually all forms of life and in cell secretions, serum and other body fluids, connective tissues, and in cell membranes. Glycans make up 10–60% of the molecular weight of glycoproteins. The diverse biological functions that these macromolecules perform include acting as energy sources, as structural components, as key elements in enzymatic catalysis, hormonal control, immunological protection, ion-transport, blood clotting and lubrication, in various molecular recognition processes including bacterial and viral infections, cell adhesion in inflammation and metastasis, differentiation, development and many other intercellular interactions and signal transduction events [3–5].

Oligosaccharides were regarded as compounds completely lacking biological specificity. However, the discovery of the role of protein bound saccharides in biological recognition changed this view. The large structural diversity of protein conjugated glycans, complexity of the biosynthetic pathways, tissue specific and developmentally controlled expression, and plasticity of the structure of complex sugars of glycoproteins in response to pathological conditions lends further support to their importance. Glycans in glycoproteins have diverse biological roles, therefore becoming one of the main topics for discussion. Since the majority of the candidate therapeutic proteins are glycosylated and produced by recombinant DNA technology, alternate systems to express them and characterization of the respective glycan structures have been discussed.

2 Site and Events of Glycosylation in Eukaryotic Cells

The majority of secretory as well as plasma membrane proteins and lysosomal enzymes are glycosylated and have a common biosynthetic origin on the rough ER [6]. Nascent proteins are translocated into the cisternal space of the ER where the signal peptide is cleaved, initial cotranslational folding and formation of disulfide bonds occurs, addition and initial processing of high Man *N*-linked 14-saccharide core unit ($Glc_3Man_9Gn_2$) takes place, and finally, for multimeric proteins, oligomerization or subunit assembly is attained before these become competent to be transported out of the ER. The semiprocessed proteins are translocated to the Golgi apparatus, further processed, and then either translocated to the cell surface or packaged into secretory vesicles for secretion [7, 8].

2.1 Entry in the Secretory Route

The majority of proteins to be targeted to the secretory route are initially synthesized as precursors containing a hydrophobic signal being either a cleavable *N*-terminal peptide [9] or a noncleavable internal sequence located near the *N*-terminus [10]. Signal peptides of some proteins have been shown to mediate efficient secretion in heterologous environments.

Protein translocation across the ER membrane can occur by two pathways. Most secretory proteins are cotranslationally translocated in contrast to posttranslational translocation destined for other cellular compartments. Transport of precursor proteins across ER membranes occurs most likely through a protein-conducting channel [11]. The signal recognition particle (SRP) and nascent-polypeptide-associated complex (NAC), a nonribosomal factor, together provide for fidelity in protein targeting to ER and serve as a versatile targeting chaperone team [12, 13]. Details of the mechanism of protein translocation across membrane are beyond the scope of this review and are not discussed. However readers are advised to see the current literature [14, 15].

2.2 Early Modifications of Protein

The cleavage of the signal peptide by signal peptidase occurs rapidly soon after the translocation of protein across the membrane of rough ER. Prokaryotic signal peptides can be released by eukaryotic signal peptidases and vice versa. Protein folding is facilitated by the resident proteins of ER; some of them are enzymes responsible for disulfide bond formation, isomerization of peptide bonds and glycosylation. Disulfide bond formation is a major rate-limiting step in protein folding, which involves the enzyme protein disulfide isomerase (PDI). The PDI essentially unscrambles intermediates in the protein folding pathway with non-native disulfide bonds [16]. Similarly, *cis-trans* isomerization of prolylpeptide bonds is catalyzed by peptidyl-prolyl *cis-trans* isomerase [17]. Other resident proteins are the classes of molecular chaperones, which appear to stabilize protein folding intermediates, prevent competitive aggregation interactions, or promote correct folding [18]. They somehow recognize the kinetically trapped intermediate states of proteins, randomly disrupting and releasing them in less folded states [19]. Their interiors provide a sticky hydrophobic surface that competes with intrachain hydrophobic collapse and helps to pull apart an incorrectly folded protein, allowing opportunities to find pathways leading to the stable native state. Chaperones appear to act sequentially in protein folding pathways.

That folding begins cotranslationally is known for some proteins including IgG, serum albumin and has been suggested for globin molecule [20]. Again, in influenza hemagglutinin, disulfide bond formation and generation of conformational epitopes are cotranslational [21]. After initial modification in ER, some proteins are retained (resident ER proteins) but the majority are exported to final destinations within and outside the cell. Export from ER involves incorporation into transport vesicles that fuse with the next compartment along the secretory pathway. The pathway consists of a series of membrane-bound organelles between which proteins move in a vectorial manner. Specific signals have been identified for the retention/retrieval of ER proteins. Retrieval of soluble ER resident proteins from Golgi is mediated through recognition of amino acid sequence 'Lys-Asp-Glu-Leu' by a specific receptor [22], while for transmembrane proteins, the intracytoplasmic dibasic motif plays a similar role [23].

Localization signals of Golgi GlycTs are more complex, and seem conformation dependent. Sequences in the cytoplasmic tails, transmembrane regions and in luminal regions are important for active Golgi retention [24]. One mechanism for retention is based on protein oligomerization in the environment of a specific Golgi cisterna, as in the case of α-2,6-ST. The oligomer is a disulfide bonded dimer of the enzyme and is catalytically inactive as active site Cys residues participate in dimerization. Under conditions of inadequate supply of donors and/or substrate, the reactive sulfhydryls in the catalytic domain of one molecule may interact with those of another. Only those that are continuously supplied with donor and substrate resist disulfide bond formation and remain active. The disulfide bonded dimer acts primarily as a Gal-specific lectin in the Golgi to retain unsialylated molecules and pass them off to the active ST for sialylation [25].

Proteins that do not reside in ER exit at specialized regions adjacent to the Golgi. Within the Golgi, different subcompartments are distinguishable (cis-, medial- and trans-Golgi) on the basis of a distinct content of glycan processing enzymes. Adjacent to the trans-Golgi cisternae is a network of tubules located with coated buds and vesicles that is called the trans-Golgi network (TGN) where sorting of proteins destined to specific intracellular compartments or cell surface domains from constitutively secreted proteins occurs [26]. The secretory pathway involves vesicular transfer to the plasma membrane followed by the secretory event, that is exocytotic discharge of vesicle contents [27].

2.3 Asparagine-Linked Glycosylation

A number of different types of protein glycosylations described in the literature [28, 29] are shown in Table 1. *N*-Glycans are characterized by a β-glycosidic linkage between a Gn residue and the δ amide *N* of an Asn residue. Though there are many different Asn-glycans, the common feature is the presence of a pentasaccharide (Man_3Gn_2) core because they all arise from the same biosynthetic precursor lipid-like oligosaccharide that is transferred to the nascent peptide chains. The core can have a Fuc attached to the Gn that is linked to the Asn, and can possess a bisecting Gn residue attached to the central Man of the core. The *N*-linked glycans fall into three main subgroups as shown in Fig. 1. These are as follows.

1. High-Man structures contain 2–6 additional Man residues at the two terminating Man residues of the core and vary in the number, position and degree of phosphorylation or sulfation of Man residues.
2. Complex-type structures have 2–4 lactosamine (Gal β1,4-Gn) units distributed over the two outer Man residues of the core forming bi-, tri-, or tetra-antennary structures. Each of the arms can terminate in sialic acid forming sialyllactosamine. Sulfated lactosamine has been found to substitute sialyllactosamine in some cases, such as the glycoprotein hormones LH, FSH and TSH. Other complex-type structures contain polylactosamine in their outer chains, as in erythroglycan.

Table 1. Classification of glycoproteins and nature of linkage

Type of glycosylation	Occurrence	Glycan - peptide linkage		Anomeric type	Stability to acid	Stability to alkali
		Monosaccharide	Amino acid			
N-glycosylation	Widespread	GlcNAc	Asn	β	+	+
O-glycosylation						
Mucin	Mucin, blood group, fetuin, antifreeze glycoproteins	GalNAc	Ser/Thr	α	+	–
Intracellular	Nucleus and cytosolic	GlcNAc	Ser/Thr	β		
Proteoglycan	Proteoglycans	Xyl core	Ser	β	+	–
Collagen	Collagens	Gal	OH-Lys	β	++	++
Clotting factor	Factor IX	Fuc/Glc core	Ser			
Fungal	Yeast and Fungal glycoproteins	Man	Ser/Thr			
Plant	Plant cell walls, earthworm, cuticle	Ara Gal	OH-Pro Ser	β α	– +	 –
GPI anchore	Cell surface receptors					

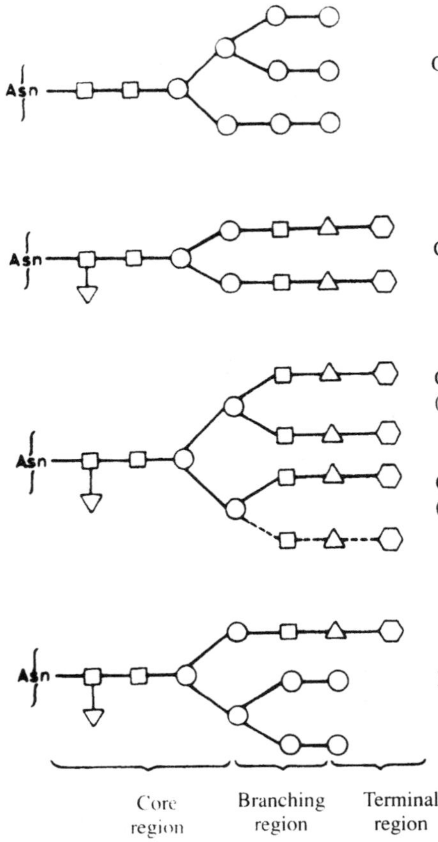

Fig. 1. The basic forms of common *N*-linked glycans. The *N*-linked glycans posses three regions of oligosaccharides - core, branching and terminal. All glycans have common core of two *N*-acetylglucosamine residues and three mannose residues. *N*-linked glycans are primarily depend on the branching pattern and the type of monosaccharides present in the branching and terminal regions. The branches are distributed over the two terminating core mannose residues. The complex type *N*-linked glycans can have two to four antennae. In complex and hybrid type *N*-glycans, the antennae usually terminate in sialic acid or galactose. Sialylation adds the greatest degree of microheterogeneity, however variable core fucosylation also adds heterogeneity to the glycan. □-GlcNAc (Gn); ▽-Fuc; ○-Man; △-Gal; ⬡-NueAc (SA)

3. Hybrid-type structures combine the structural features of both the high Man and the complex-type.

2.3.1
Biosynthesis and Processing

Glycosylation begins in the rough ER and proceeds as glycoproteins migrate through the Golgi to their final destination. Asn-glycosylation is a cotranslational event and occurs as the polypeptide is being transferred into the ER, while still in an unfolded state [30]. Initial steps in this pathway, which appear to be generally conserved, involve the synthesis in ER membrane of a dolichol linked precursor glycan and its transfer to an Asn residue of the growing peptide chain. The major steps in phosphodolichol pathway of protein *N*-glycosylation have been reviewed [31]. The principal pathway for biosynthesis of the dolichol-linked oligosaccharides is shown in Fig. 2. The oligosaccharyl transferase (OST) is integral to ER membrane, with active site of the enzyme residing near the membrane on the lumenal side, and transfer only occurs when 12–14 amino acids *C*-terminal to a sequon have been translocated into the ER lumen [33]. The glycan is added to Asn residue in the tripeptide consensus sequence Asn-X-Ser/Thr (X = any amino acid) and presence of Thr rather than Ser at the hydroxy position favors efficient glycosylation [34]. The OH group of the Thr/Ser has a role in explaining the unique reactivity of this Asn with the DolPP sugar derivative. Secondary structure prediction analysis suggested that these tripeptide consensus sequences have a high probability of occurring in a β-turn or other loop structures, which could serve as recognition sites for enzymes involved in glycosylation [35]. Analysis of occurrence of amino acids at positions around the glycosylated Asn (given the position zero) indicates that Pro is rarely allowed at position +3 and is never observed at position +1 [36]. Asp, Glu, Leu and Trp at position +1 are also associated with inefficient glycosylation [37]. This position is preferentially occupied by non-bulky amino acids such as Gly, Val and Ala.

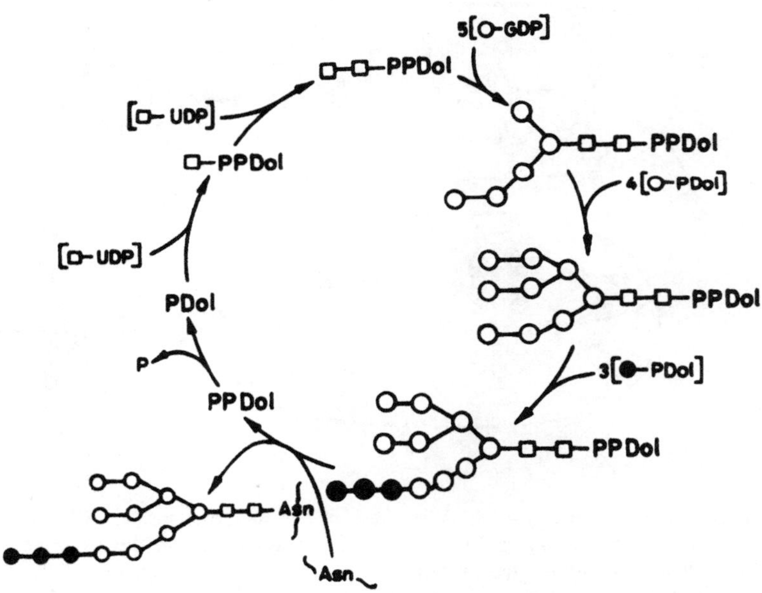

Fig. 2. Pathway of biosynthesis of the dolichol-linked oligosaccharide. □-UDP: GlcNAc-Uridine diphosphate; □-PPDol: GlcNAc-Pyrophosphatedolichol; ○-GDP: Man-Guanosine diphosphate; ○-Pdol: Man-Phosphodolichol; ●-Pdol: Glc-Phosphodolichol. (adopted from [32])

The consensus sequence Asn-X-Ser/Thr is a necessary, but not sufficient, requirement for addition of Asn-linked glycans. Less than half of the known tripeptide sequences in secreted proteins are only glycosylated, and this may be due to differences in accessibility of a sequon in a protein and to OST and the folding of the nascent protein chain [38]. The conformation and rate of translation of nascent polypeptide too influence the frequency of sequon utilization. Thus, preventing cotranslational disulfide bond formation in ER in presence of dithiothreitol (DTT) leads to complete glycosylation of a sequon in tPA that otherwise undergoes variable glycosylation in untreated cells. This shows that

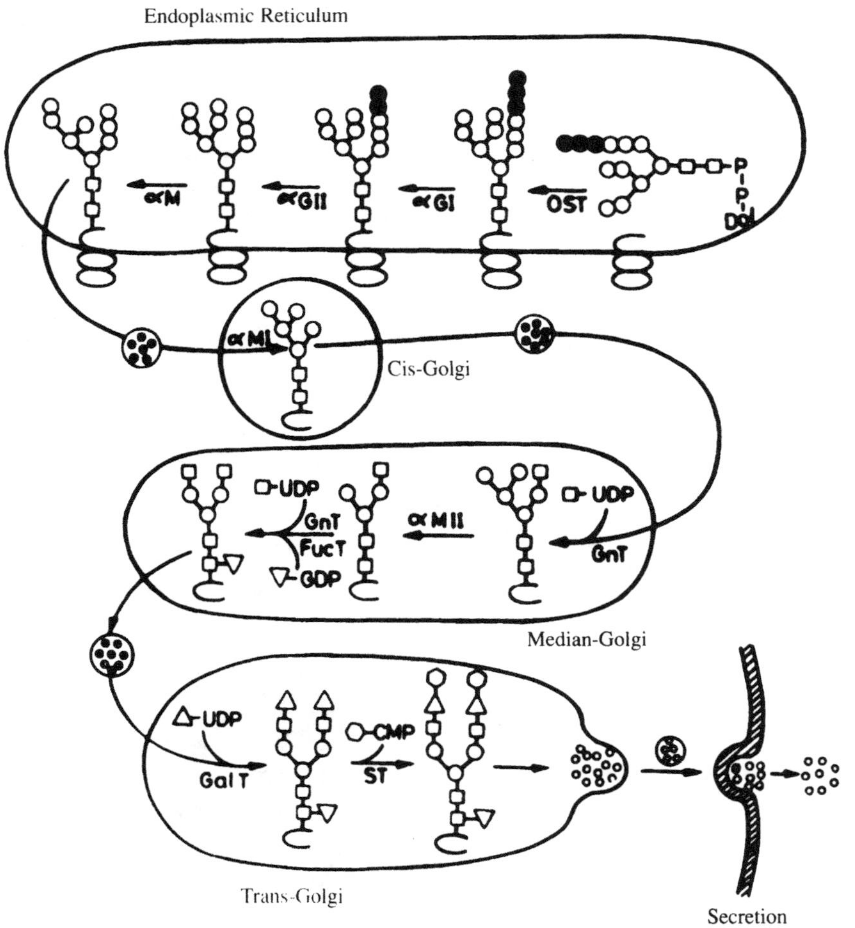

Fig. 3. Representative pathway for the biosynthesis of Asn-linked oligosaccharide. Biosynthesis starts in the rough endoplasmic reticulum and proceeds through the *cis*-, medial-, and *trans*-Golgi apparatus, before glycoprotein is secreted. Some of the enzymes involved are: oligosaccharyl transferase (OST); α-glucosidase I & II (αG I & II); ER-mannosidase (αM); α-mannosidase I & II (αM I & II); *N*-acetylglucosaminyl transferase I & II (GnT I & II); fucosyl transferase (FucT); galactosyl transferase (GalT); sialyl transferase (ST). □-Gn; ○-Man; ●-Glc; ▽-Fuc; △-Gal; ⬡-SA (adopted from [40])

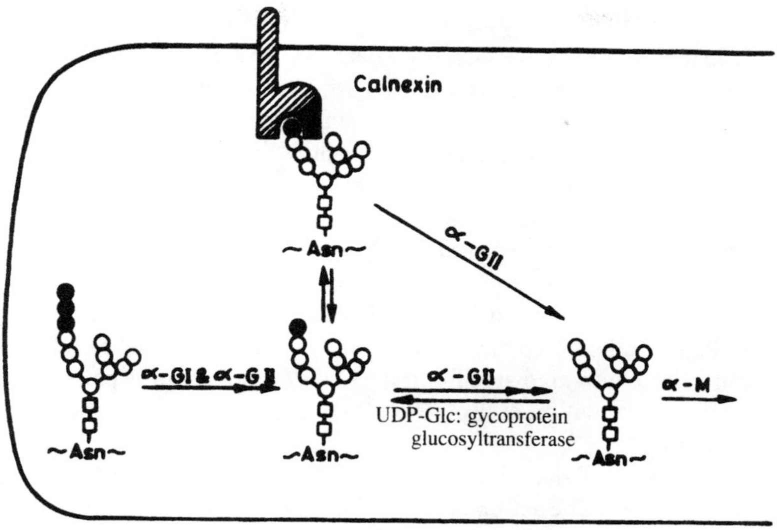

Fig. 4. A model for the involvement of calnexin in quality control mechanism in ER. Association and disengagement of calnexin depend on the folding status of the semiprocessed glycoprotein. Calnexin recognizes partially trimmed monoglucosylated glycans, generated either by glucose trimming or by reglucosylation pathways. Symbols and abbreviations are same as Fig. 3 (adopted from [45])

folding and disulfide bond formation may determine extent of core *N*-glycosylation [39]. Analysis of glycosylation site-occupancy has revealed that glycosylation of potential target sequons is more likely to occur near the *N*- than the *C*-terminus of a protein [36].

The glycan that is transferred to the protein is a substrate for a variety of processing α-Gs and GlycTs to give mature high Man or complex-type chains as given in Fig. 3 [40]. The glycan is processed in terms of removal of three Glc residues by stepwise action of α-G I and α-G II. The Glc trimming affects glycoprotein exit from the ER. A quality control procedure ensures that proteins which are incompletely folded, or incorrectly oligomerized, are not transported out of the ER so long as they have not acquired "export competent conformation" [41]. Quality control and degradation depend on glycosylation as deglycosylated species are stably retained in ER, as observed for Na, K-ATPase β-subunit [42]. The role of Glc residues in protein folding and quality control has been clarified by the identification of two lectin-like proteins in the ER – calnexin and calreticulin; the latter is lumenal while the former is a transmembrane molecular chaperone protein [43, 44]. Calnexin is believed to recognize partially trimmed monoglucosylated ($Glc_1Man_9Gn_2$) glycans on newly synthesized glycoproteins and detain them in the ER until properly folded. Monoglucosylated chains may be generated via two mechanisms – cotranslational processing of immature $Glc_3Man_9Gn_2$ glycans (the Glc trimming pathway) by α-G I and α-G II, or reglucosylation of fully trimmed $Man_{8-9}Gn_2$ glycans by lumenal UDP-Glc:glycoprotein glucosyltransferase (reglucosylation or salvage pathway). A model proposed for the involvement of calnexin in quality control of influenza hemagglutinin (HA) is shown in Fig. 4. The glucosyltransferase preferentially acts on unfolded proteins and functions as a major sensor for incompletely folded proteins in ER. During folding, a glucopeptide cycle is formed between fully trimmed and monoglucosylated glycans. Depending upon folding status, the glycoprotein enters into the de- and reglucosylation cycle. Properly folded protein escapes the reglucosylation step, and the deglucosylated form is liberated from the calnexin anchor for subsequent processing. Proteins that fail to attain the correct conformation or have assembled into non-native aggregates are retained in ER, and degraded by ubiquitin-proteasome system [46]. It may be noted that the calnexin-calreticulin mediated folding pathway is not the only pathway to be followed in the glycosylation of proteins; other chaperones seem to be involved for protein folding.

Furthermore, α-M efficiently removes a single terminal Man residue to generate Man_8Gn_2 in the ER. The action of ER α-M perhaps alters the conformation of the glycoprotein, making it more susceptible to the Golgi α-M I. The concerted efforts of ER and Golgi α-Ms may be required for fine tuning mechanism to produce surface glycoproteins with particular assortments of high Man type chains [47]. In Golgi, Man chains are processed in two steps by the action of α-M I, active in the cis-Golgi and α-M II, active in the medial-Golgi, resulting in the removal of additional five Man to yield Man_3Gn_2 [48]. The processing of Man chains is obligatory for the formation of complex-type glycans. In medial Golgi, GnT I inserts Gn residue to α1,3-linked Man followed by the cleavage of two terminal Man by the action of α-M II. Addition of another Gn to the α1,6-linked Man by GnT-II takes place in the medial Golgi. In case of mammalian cells, only occasionally, but in plant cells generally, Fuc residue is also incorporated at this stage in the medial-Golgi by the action of α-FT. The termination of glycan chain occurs in the trans-Golgi by the incorporation of Gal and SA with the help of GalT and ST, respectively. Thus, a complex-biantennary *N*-glycan is synthesized before secretion (Fig. 3). In different cell types, diversity in structural features of *N*-linked oligosaccharides is obtained, although these are not included in Fig. 3. As an example, GnT III catalyzes the addition of Gn in β1,4-linkage to β-linked Man of the core producing a bisecting Gn residue, and controls the branching patterns in vivo [49]. Amongst other terminal glycosylation, incorporation of GalNAc, sulfation, and *O*-acetylation of SAs are important, also occurring in the trans-Golgi, which is again cell specific. The turnover of terminal glycans is much faster than that of the protein molecule, and the core sugars exhibit a turnover rate similar to that of the protein [50].

2.3.2
Factors Regulating Asn-Linked Glycosylation

Despite the fact that the *N*-linked glycans are derived from the same precursor, with few exceptions each glycosylated site in a glycoprotein is associated with

Table 2. Factors regulating Asn-linked glycosylation

1.	Array and activities of Golgi GlycTs, including GnT, GalT, FucT, ST
2.	Polypeptide structure and physical accessibility of oligosaccharides to enzymes
3.	Oligomerization of glycoprotein subunits
4.	Order in which the glycoprotein encounters the processing glycosidases and GlycTs
5.	Availability of dolichol and dolichol-linked donors
6.	Transit time of glycoprotein in ER and Golgi
7.	Competition between two or more GlycTs for a common substrate

several different glycan structures (site microheterogeneity). Some factors that contribute to the regulation for the biosynthesis of Asn-linked glycans are given in Table 2. The polypeptide structure and overall conformation strongly influence the type of modification that glycan chains undergo. Glycosylation sites committed to becoming complex type structures are relatively more exposed. The nature of glycoforms found in a glycoprotein is species and tissue specific, and cell development and differentiation are accompanied by alteration in glycosylation patterns [51]. Since this structural variation is confined to the terminal glycosylation sequences, the synthesis must be highly regulated at the processing level by Golgi GlycTs [52]. The number of glycosidases and GlycTs involved in the synthesis of both the *N*- and *O*-linked glycans is estimated to be over 100 [53]. In general, each enzyme is specific for the structure of an acceptor oligosaccharide and adds a monosaccharide in a particular linkage at a precise location. The high level of specificity displayed by GlycTs allows them to synthesize complex structures with high degree of fidelity.

2.4 The Biosynthesis of Ser/Thr-Linked Glycans

The *O*-linked glycoproteins contain an α-glycosidic linkage between a GalNAc and the hydroxyl group of a Ser or Thr residue of peptide chains. The chain elongation requires the sequential addition of monosaccharide residues. There are eight core structures that have been identified in *O*-linked glycan [54], side chains to which may be branched and have varying degrees of complexity. A common feature of all *O*-linked glycans is the presence of nonreducing terminal α-linked sugars (SA and L-Fuc) and the absence of Man and Glc residues. The structure of mucine-type oligosaccharide is shown in Fig. 5. *O*-Glycosylation is a post-translational event, taking place in the Golgi after *N*-glycosylation, folding and oligomerization, and is limited to residues present at protein surface. Rules that govern placement and structure of *O*-glycans on glycoproteins remain unclear and little is known about factors that start *O*-glycosylation steps.

A cumulative specificity model, deduced from the amino acid sequences surrounding 90 Ser and 106 Thr *O*-glycosylation sites, has been inferred for the acceptor substrate specificity of GalNAcT, catalyzing the first committed step of *O*-glycosylation. The specificity is consistent with the existence of an extended

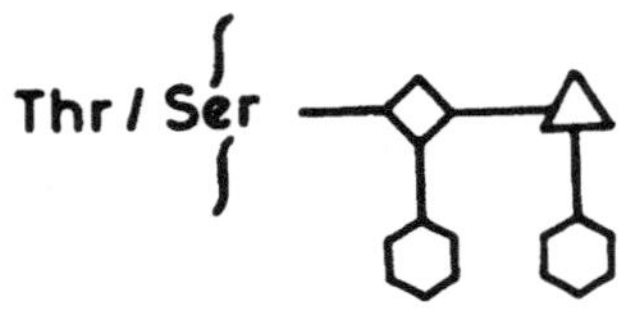

Fig. 5. The mucine type *O*-linked glycans: ◇-GalNAc; △-Gal; ⬡-SA

site composed of nine subsites with the acceptor Ser/Thr in the centre. The model postulates independent interactions of the nine amino acid moieties with their respective binding sites [55]. However, no consensus sequence has emerged due to the broad range of residues that the binding site of GalNAcT can accommodate, and due to the existence of multiple isoforms of GalNAcT with overlapping specificities [56]. The distribution of charged amino acids flanking the *O*-glycosylation site can have a large influence on glycosylation, with position −1 relative to the glycosylation site being particularly sensitive. A combination of acidic residues at positions −1 and +3 almost completely eliminates glycosylation. An amino acid change resulting in de novo attachment of *O*-linked glycan on glycoprotein hormone common α-subunit has been reported [57].

There are seven different types of *O*-glycans available in nature [29], amino acids to which glycosylation occur and corresponding *O*-linked sugars have been shown in Table 1. Analysis of mucin type glycosylation revealed that Pro occurs at increased frequency at positions −1 and +3 relative to the glycosylation site [58]. The acceptor sequence context for *O*-glycosylation of Ser was found to differ from that of Thr, and showed a high abundance of Pro, Ser and Thr. In general, the *O*-glycosylation sites are found to cluster and to have a high abundance in the *N*-terminus of the protein. The sites are also found to have an increased preference for different classes of β-turns. Pro in positions −1 and +1 is speculated to function as 'gating' residue favouring *O*- and inhibiting *N*-glycosylation [29].

2.5 Glucosylphosphatidylinositol (GPI) Anchors

Anchoring of surface membrane proteins (e.g. receptors) via GPI anchors is a major means in eukaryotic cells. All GPI anchors analysed contain a common glycan core (Man α1-2Man α1-6Man α1-4GlcNH). This may be further processed in cell and protein specific manner. Nascent proteins destined to be GPI-anchored contain, besides the amino terminal signal peptide that is typical of proteins processed in ER, a second hydrophobic peptide at their *C*-terminus which is also removed during processing. The GPI moiety is linked to what had been an internal sequence in the nascent protein. The subject matter of assembly of GPI anchors [59] and GPI anchored membrane proteins [60] have been reviewed recently.

3
Functions

The observed large diversity in the structure of Asn-linked glycans is central to the hypothesis that these are important in biological functions [3, 7, 28, 61, 62]. It is clear that there is no unifying single specific role of glycans and in some cases they may not even have any function at all and be completely replaceable. The effect of carbohydrate on the physicochemical properties of glycoproteins, such as viscosity, solubility, isoelectric pH, degree of hydration, and other structural roles have been known for some time [2]. Studies on the native glycosylated, carbohydrate-depleted, and recombinant nonglycosylated proteins have revealed such effects as stabilization of protein conformation, protection from proteolysis and enhancement in solubility [63, 64].

The first clear demonstration of the functional significance of carbohydrate was in the blood group substances, where the immunological specificity was found to depend on monosaccharides or short glycan chains. A major breakthrough occurred when it was discovered that the removal of SA resulted in rapid clearance of glycoproteins from circulation [65]. Since then, the role of glycans in a variety of other functions has been reported. Glycoconjugates play roles in many cell-cell recognition processes, including metastasis and inflammation. Glycan structures both mediate and modulate cell-cell and cell-matrix interactions [66].

3.1
Protein Folding and Conformation

Bound carbohydrates can influence protein structure and the effects depend upon type of sugar, linkage, stereochemistry, size of the bound saccharide and characteristics of the protein. The conformational effects of protein glycosylation have been studied using various spectroscopic techniques [67–69]. In one such study using time-resolved fluorescence energy transfer (FET), it was suggested that cotranslational glycosylation can trigger the timely formation of structural nucleation elements, prevent aggregation of partially structured chains by improving solubility, and generally assist in protein folding [69].

The role of *N*-glycosylation and disulfide bonds in the folding of proteins has been studied extensively [70, 71]. Inhibition of core glycosylation with inhibitors (e.g. tunicamycin) or site-directed mutagenesis leads to misfolding, aggregation and degradation of proteins retained in ER [7]. Absence of *N*-glycosylation leads to impaired lipoprotein lipase secretion and accumulation of inactive protein in ER [72]. Unglycosylated rabbit testicular angiotensin-converting enzyme (ACE T) is inactive and rapidly degraded intracellularly. However, allowing glycosylation only at the first or second site (out of five *N*-glycosylation sites) as counted from the *N*-terminus was sufficient for normal synthesis and processing of active ACE T [73].

Studies with simian viral hemagglutinin neuraminidase and yeast acid phosphatase suggest that *N*-glycans are needed for proper folding of glycoproteins

[74]. Similarly, viral spike glycoproteins depend on *N*-glycosylation for proper folding and transport to cell surface. Those with an absolute requirement for glycosylation include the D-glycoprotein of herpes simplex virus, the sendai virus glycoprotein, the hemagglutinin of influenza X31 and the vesicular stomatitis virus-Indiana glycoprotein [75]. The initial step in human immuno-deficiency virus (HIV) infection involves the binding of gp120 to the cell surface molecule CD4, and *N*-glycans are found to be essential for generation of proper conformation of gp120 to provide a CD4 binding site [76].

There are two potential *N*-glycosylation sites in human IFNγ. Proper glycosylation is found to be essential for dimerization, efficient secretion and biological activity of IFNγ [77]. Nonglycosylated IFNγ exhibited only 50% of the antiviral activity of the native molecule. It has been found that glycosylation at Asn^{25} is essential in the folding and dimerization of newly synthesized molecules, which also provide resistance against common cellular proteases [78].

Again, Matzuk and Boime have shown that the assembly of *N*-glycosylation mutants of hCGβ-subunit with its counter subunit is decreased due to an alteration in folding [79]. Out of two *N*-glycans (Asn^{30} and Asn^{13}), Asn^{30} glycan is found to be more important for efficient folding of hCGβ. However, once hCGβ folds correctly, the *N*-glycans are no longer involved in its assembly with α-subunit [80]. Furthermore, the composition of *N*-linked glycans of hCGβ did not change protein folding, as substrates with high-Man glycans fold as efficiently as substrates containing complex glycans. Inefficient folding of hCGβ lacking both *N*-glycans correlated with the slow formation of last three disulfide bonds in the hCGβ folding pathway. However, coexpression of hCGα gene enhanced folding and formation of disulfide bonds of hCGβ lacking in *N*-glycans [81]. For hCGα subunit, the Gn residue at Asn^{78} seems to be crucial for folding and maintenance of stability. In an interesting observation it has been found that minor modifications in *N*-glycans of placental hCGα prevent combining with β-subunits. These modifications include a higher degree of glycan branching, as evidenced by larger amounts of Fuc, SA, Gal and Gn in hCGα subunit [82]. In a comparative study of in vitro folding of glycosylated and nonglycosylated hCGβ, it reveals that most of the nonglycosylated protein folded into biologically inactive form [unpublished findings].

3.2
Protein Stabilization and Structural Integrity

Glycans stabilize glycoprotein structure, decrease global dynamic fluctuations, and prevent degradation and protect proteins from the unfolding state. The terminal sugar residues (antennae) often serve as recognition markers and modulate biological functions such as cell-adhesion, cell-extracellular matrix interactions or protein clearance from circulation through surface interactions. The core sugar residues function primarily as supporting structures between the polypeptide and the outer sugar residues. The core sugar residues are necessary and sufficient for structural integrity and in maintaining a functional polypeptide structure. In factor-X activator of Russell's viper venom, the core glycans are involved for maintaining integrity of secondary structure and not the peripheral glycans [83].

The stabilizing effect can even be brought about by a single sugar residue [2]. The attached glycans can stabilize protein conformation by forming hydrogen bonds or having other hydrophobic interactions with the polypeptide backbone. Circular dichroism (CD) spectroscopy of model glycopeptides suggests that bound glycans interact directly by hydrogen bonding with the peptide backbone to stabilize a particular structure [84]. In human corticosteroid binding protein it has been presumed that the glycan interacts with a tryptophan residue in the protein to create a stable steroid binding site [85].

A high degree of glycosylation induces a well-defined saccharide conformation and an extended peptide backbone structure. The radius of gyration, a measure of the statistical average distance of the end of the chain to its centre, of mucin chain is 2.5 to 3-fold larger than that of a denatured polypeptide chain of equal number of amino acid residue [86]. The glycans tend to stiffen the polypeptide backbone and the structure of mucin approaches that of a rigid rod. This effect of bound carbohydrate on conformational stability is important for orientation and function of membrane bound glycoproteins. In an another study with five model glycoproteins from various sources, differential scanning microcalorimetry and CD spectroscopy indicated that glycans have an apparent stabilizing effect on conformation and enhance thermal stability [87]. Thus, glycosylated enzymes expressed in *S. cerevisiae* are more heat stable than their unglycosylated forms expressed in *E. coli* [88].

The stereochemical features of the *N*-glycosidic linkage between the first Gn and Asn, important for the orientation of glycan chains, have been statistically analyzed employing 44 different glycosylation sites belonging to 26 glycoproteins [89]. It was found that *N*-glycosylation does not significantly change the rotamer distribution for the Asn side chain as compared to nonglycosylated Asn. Carbohydrates have a high hydrogen bonding potential, but bonding between Gn and peptide is infrequent and Gn shows a general tendency to extend into the solvent. Freedom of rotation about the glycosidic bonds and solvent exposed nature accounts for the flexibility of attached glycans. However, steric limitations often restrict the rotation about some of the linkages [90].

In most X-ray crystal structures of glycoproteins, only the first 1–4 sugar residues most proximal to the glycosylation site are defined by proper electron density, except in few cases where longer fragments of carbohydrates have been resolved [91]. Hydrogen bonds and hydrophobic interactions between polypeptide and attached glycan have been observed. Study of crystal structure of glucose oxidase showed that the *N*-linked Man residues form hydrogen bonds with the backbone *N* and the carbonyl *O* of Glu [92]. Computer simulation of molecular dynamics of ribonuclease B also revealed possible hydrogen bonding of *N* of Lys side chain with the ring *O* and the hydroxymethyl group of glucosamine [93].

The partial NMR spectroscopic studies of intact glycoproteins so far reported, indicates that the core Gn of *N*-glycan at amino-terminal adhesion domain of human CD2 interacts with the polypeptide part of the molecule. As a result the mobility of the proximal glycan residues is restricted. The glycan counterbalances an unfavourable cluster of positive charges from surface Lys, through hydrogen bonds and van der Waals contacts, and thus plays a role in stabilizing the native receptor structure [94]. The solution structure of CD58 revealed that the *N*-glycan, located opposite to the binding site, is not directly involved in ligand binding, and a single Gn residue stabilizes the receptor structure [95].

Hydrogen-deuterium exchange kinetics reveals that, while glycans had little overall effect on the three dimensional structure of the glycoprotein, there was a global decrease in dynamic fluctuations due to hydrophobic and hydrogen bond interactions which correlated with an increase in stability by ~1 Kcal mol^{-1} [63, 96]. The same glycan on different proteins may have quite different effects depending on the orientation with respect to the polypeptide [97]. Also, different glycoforms of a protein may display quite different orientations of the glycan with respect to the protein, thus conferring different conformations [94]. For example, the conserved complex glycan on each heavy chain of IgG Fc C_H2 domain occupies the interstitial space between the domains and stabilizes Fc hinge conformation. The antenna of each glycan, in particular the terminal Gal, interacts with hydrophobic and polar residues on the domain surface. Loss of the two terminal Gal, as in patients with rheumatoid arthritis, results in a loss of interaction with the C_H2 domain surface. This displaces and exposes the *N*-glycan, giving them the potential to be recognized by endogenous Man-binding proteins [98].

3.3 Receptor Functioning

The exact role of *N*-glycans in the function of glycoprotein receptors is varied. Receptor glycosylation is generally thought to mediate correct folding and insertion of the protein into the extracellular membrane, as in the case of FSH receptor [99]. Glycosylation is also thought to be important in ligand binding in some receptors, such as somatostatin receptor [100], vasoactive intestinal peptide (VIP) receptor [101], and in the functioning of rhodopsin, the dim-light photoreceptor of the rod cells [102], CD2 receptor [103], and in the α-subunit of the human granulocyte macrophage colony stimulating factor (GM-CSF) receptor [104].

In human Ca^{2+} receptor [105] and in FSH receptor [99] *N*-glycosylation is essential for cell surface expression. Similarly, for the mature LH/CG receptor to reach the cell surface, post-translational processing of high Man glycans to complex glycans is essential [106]. Inhibition of *N*-glycosylation prevents cell surface presentation of VIP receptor as in FSH receptor [101]. Changes in cell surface expression likely reflect abnormal folding of the nonglycosylated receptor protein or decreased stability or transport. In some cases, though the receptor contains more than one glycosylation site, glycans at one of the sites alone is adequate for the proposed role as in FSH receptor [99], human transferrin receptor [107] and rhodopsin [102].

There have been some reports where no or only a minor role of attached glycans was detected. For example, though the rat lutropin receptor is heavily glycosylated, *N*-linked carbohydrates are not absolutely required for proper folding into a form capable of binding hormone and signaling [108]. In rat angiotensin II type-2 receptor, *N*-glycans have a minor contribution to the ligand binding/affinity of the receptor and are not essential for targeting and expression of the receptor at cell surface [109].

3.4 Intracellular Trafficking

Glycans are modified for direct targeting proteins to specific locations within and outside the cell. The primary role of mannose-6-P receptor (MPR) in the Golgi is to target more than 40 different hydrolytic enzymes to lysosomes by interaction with terminal Man-6-P residues. Most mammalian cells contain two distinct transmembrane glycoprotein MPRs, and both mediate transfer of newly synthesized lysosomal proteins from trans-Golgi to endosomal compartment [110]. Similar to Man-6-P modification of lysosomal enzymes, the possibility exists that plasma membrane and secretory proteins also employ *N*-glycans as sorting signals for transport. If *N*-linked glycans are used as sorting signals in biosynthetic trafficking, the core residues should be considered as possible recognition targets [3]. In spite of the postulate for the requirement of sorting signals for trafficking beyond ER, this is an issue that being debated, and the final answer is not distinct [111]. It has recently been observed that glycoprotein sorting is related to specific structures of glycans, as for example bisecting Gn acts as a negative sorting signal for cell surface glycoproteins [112].

3.5 Cellular Recognition and Adhesion

Cell surface oligosaccharides are central to cell-cell communication, cell-adhesion, infection, differentiation and development. The potential for enormous variation in glycan structure, permitting fine tuning of recognition determinants, is an appealing feature of carbohydrate mediated cellular adhesion. Furthermore, the large size of glycans allows them to cover functional sites and hydrophobic patches in proteins, and modulate protein functions [2]. Protein-carbohydrate interactions are employed in a number of instances of cellular adhesion, in systems as varied as binding of pathogenic bacteria to animal cells [113], neuronal development [114], lymphocyte homing [115], immune recognition [116] and cellular migration during embryogenesis [117]. It was proposed that adhesion of these cells was caused by the interaction between glycans in the membranes of one cell with the GlycT present in the membranes of adjacent cells.

The binding of viruses to host cells involves the viral hemagglutinins and the host cell surface glycoproteins [118]. Similarly, other cell surface carbohydrates act as receptors for pathogens and toxins. Cell surface glycoconjugate containing *N*-acetyl-9-*O*-acetylneuraminic acid is believed to be involved in the attachment of influenza C virus [119]. Again, sialic acid in glycophorin, a glycoprotein in the membranes of red blood cells, has a role in the binding of viruses to receptors. Removal of negative charge by chemical amidation blocks association of encephalomyocarditis virus to receptors, suggesting a direct role for SA in attachment [120].

Carbohydrate mediated cell-adhesion is involved in metastasis and can be initiated by tissue injury or infection. One of the initial events in host response to inflammation is the ordered migration of leukocytes to inflammatory sites, initiated by relatively low affinity interactions that allow for leukocytes to roll along the inner lining of blood vessels, called the endothelium. Finally, the recognition is mediated in most regulated manners by glycoprotein E-selectin expressed on the surface of endothelial cells in response to inflammation, and a glycan displayed on the surface of neutrophils occurs [121]. The adhesion process is stimulated by signaling molecules (e.g. interleukin 1 and tumor necrosis factor) or other inflammatory factors (e.g. toxins, lipopolysaccharides, leukotrines) that induce the production of E-selectin. Another glycoprotein, P-selectin is expressed on thrombin activated endothelial cells. Passing leukocytes adhere to these protruding selectins, as their carbohydrate coat contains complementary ligands. The nature of carbohydrate ligands for the selectins appear to be complex where Fuc and SA are critical functional components [122]. Following this initial phase, tight adherence of leukocytes to the endothelial cells takes place which involves protein-protein interactions mediated by integrins on leukocytes and a protein ligand on endothelial cells. Finally, leukocytes extravasate into the underlying tissue, where they play a key role in host defense [122].

The lectin-carbohydrate recognition has also been implicated in the early development and differentiation of cells. Neural cell adhesion molecules (NCAM), are the cell surface glycoproteins which plays a key role in the development of the nervous system [123]. NCAM, with polysialic acid (PSA) appear during the early development and are believed to be involved in cellular adhesion. The negatively charged PSA, more abundant in embryonic brain than in adult brain, is implicated in reducing NCAM adhesive capability, and thus in allowing increased neurite outgrowth and cell mobility.

3.6 Receptor Binding and Signal Transduction

Glycosylation adds to the heterogeneity and signaling functions of the cytokines, including the interferons, interleukins, colony stimulating factors [124]. However, the most well studied proteins in which direct participation of the glycan moiety in their function including subunit assembly, receptor binding, function and modulation of plasma half-life has been shown are the pituitary hormones TSH, LH and FSH, and the placental hormone CG [125, 126]. These are heterodimeric in nature consisting of a common α-subunit and a unique hormone specific β-subunit. Both the subunits in all four hormones are extensively glycosylated. The integrity of the dimeric form of the hormone is essential for receptor-binding as individual subunits do not bind to the receptor and are devoid of biological activity. It is held that receptor specificity is determined by the β-subunit, but modifications to either α- or β-subunit can disrupt receptor binding [127]. In case of hCG, glycosylation in $Asn^{\alpha 52}$ has been demonstrated to be essential for maximization of in vitro bioactivity and signal transduction [128], while at the same time $Asn^{\alpha 52}$ did not show any activation of FSH receptor by hFSH [129]. In contrast, glycans at $Asn^{\alpha 52}$, and specifically the terminal SA residues, attenuates in vitro hTSH activity to a much higher degree than those at $Asn^{\alpha 78}$ or $Asn^{\beta 23}$ [130]. Deletion of both *N*-glycans on α-subunit resulted in a significant reduction of hTSH bioactivity [131].

Although structure of glycoprotein hormone receptors and hormone-receptor complexes have not been solved, several models of hormone-receptor interaction have been proposed [132, 133]. The molecular mechanism by which carbohydrates activate the receptor is believed to occur at a post-receptor binding step [125]. An indirect mechanism involving a conformational change of the hormone appears more likely than a direct interaction of glycan with the receptor. Reports on crystal structure of hCG located the glycan at position $Asn^{\alpha 52}$ to be at the dimer interface and within the proposed receptor binding region [134]. Since the hormone specific β-subunit influences common α-subunit conformation in a hormone-specific manner, differences in the spatial orientation of $Asn^{\alpha 52}$ glycan may contribute to its differential role in these hormones. The reported weak thyrotropic activity of hCG has been linked to a direct interaction with hTSH receptor. This activity of hCG increased upon desialylation in contrast to a decrease at its native receptor level [135]. Thus, the observed differences in the role of individual glycans may be related to differences in receptor structure and/or to receptor dependent differences in receptor-ligand interactions.

3.7
O-Glycosylation Functions

The lack of consensus sequence and specific inhibitors for *O*-glycosylation has hindered studies on the functions of *O*-glycans. The alterations of *O*-glycan structures often show cancer and inflammatory diseases. There are several studies indicating that *O*-glycans function as ligands for receptors modulating such diverse actions as lymphocyte trafficking, sperm-egg binding and tumor cell adhesion [136, 137]. Free swimming sperm recognize and bind to acidic glycoprotein zona pellucida (ZP3) on the egg's zona pellucida to initiate fertilization process. ZP3 from different mammals consist of a well conserved polypeptide that is differentially glycosylated. It has been shown that removal of *O*-glycans from ZP3 destroys its sperm receptor activity, whereas removal of

N-glycans or SA had no effect [138]. A Gal residue at the nonreducing terminus of these *O*-glycans is essential for binding of sperm to the zona pellucida.

The stable cell surface expressions of human interleukin-2 receptor, low density lipoprotein receptor and the major antigen envelope protein of Epstein-Barr virus depend on normal *O*-glycosylation [139]. The *O*-glycans may inhibit proteolytic degradation of proteins [28]. The role of *O*-glycan in the activity of human granulocyte colony stimulating factor (G-CSF) was suggested to be either stabilization of protein conformation or inhibition of polymerization, which would deactivate the factor [140].

Ser/Thr *O*-linked Gn is found primarily in the cytoplasm and in the nucleus. The role of *O*-Gn as regulating modification is not completely understood. *O*-Gncylation appears to be as abundant as phosphorylation, and is a highly dynamic and a regulated process [141]. *O*-Gn may have a role in modulating either the phosphorylation state or the assembly and disassembly of multimeric protein complexes in several major cellular systems including transcription, nuclear transport, and cytoskeletal organization [142].

4 Altered Glycosylation Pattern and Diseases

Inappropriate expression of GlycTs is reflected in the altered glycosylation patterns that accompany viral and chemical transformations, oncogenic events and other pathological states such as rheumatoid arthritis and cystic fibrosis. Increase in sialylation and branching of glycan structures expressed by cancer cells have been correlated with increase in metastatic capacity of these cells [143]. These changes in the glycosylation profile can be exploited in the diagnosis, as the serum level lipid-bound sialic acid (LSA) has been found to increase in most cases of metastasis diseases [144, 145].

Again, unregulated appearance of selectins and inappropriate extravasation of leukocytes results in tissue damage associated with a number of inflammatory diseases, such as rheumatoid arthritis, asthma, myocardial infarction, and acute lung injury. It is important for the therapeutic control of selectin synthesis. New anticancer agents could be developed based on the inhibition of adhesion by neutralizing antibodies and administration of competing glycans or peptides to block receptor-ligand interaction. The carbohydrate ligands of E- and P-selectin have been shown to be potentially useful for this purpose [146]. Another means to inhibit selectin activity is to identify compounds that selectively block either the signaling pathways or the transcription factors involved in induction of genes encoding the selectins [147].

The molecular basis of inherited defects is established for a number of diseases and altered glycosylation patterns have been observed to be associated with these. For example, IgG Fc fragment conserved site glycans lacking in outer arm Gal residues increase in rheumatoid arthritis, tuberculosis and Crohn's disease. There is also an Fab-specific increase in glycans bearing a bisecting Gn and a core Fuc [148].

4.1
Carbohydrate Deficient Glycoprotein Syndrome (CDGS)

Persons with CDGS exhibit abnormal glycosylation of several serum glycoproteins (α-1 antitrypsin, transferrin). It is a hereditary multisystemic glycosylation disorder with involvement of the central and peripheral nervous system leading to mental retardation and hypotonia [149]. This appears to be a highly substrate specific defect [150]. The deficiency of SA results in the presence of abnormal isoforms. Glycoprotein transport along the secretory pathway is delayed and dilation of ER indicates a retention of misfolded glycoproteins. Alternatively, CDGS could be due to defective synthesis and transfer of dolichol-linked precursors [151]. The disease is usually diagnosed by isoelectric focusing (IEF) of the proteins.

4.2
Cystic Fibrosis

A number of mutations have been found to exert their effect by specific change in a particular protein leading to altered protein folding and defective modification. Some of the human diseases for which altered protein folding and inability of the mutant protein to achieve its functional conformations are responsible include cystic fibrosis (CF) and maple syrup urine disease [152]. Cystic fibrosis is one of the most frequent inherited lethal disorders in humans and is caused by the functional absence of a plasma membrane Cl channel, designated cystic fibrosis transmembrane conductance regulator (CFTR) [153]. Clinical symptoms include both pancreatic and pulmonary insufficiencies. The vast majority of severe CF cases are linked to a single genetic lesion, deletion of a Phe codon (ΔF508). This interferes with the folding of newly synthesized CFTR polypeptides and leads to incomplete *N*-glycosylation, failure to traffic to the plasma membrane, and retention and degradation by the ER quality control mechanism.

4.3
Amyloid Diseases

Amyloidosis is a group of diverse conditions in which one of the 16 normally soluble and functional proteins self-assembles into an insoluble protease resistant β-sheet fibril form. The insoluble fibril associates with plasma and extracellular matrix proteins and proteoglycans to form neurotoxic amyloid deposits as in Alzheimer's disease [154]. It is a progressive neurodegenerative disorder of the aged and characterized by a series of structural abnormalities in the brain, including dense extracellular aggregates called senile plaques. The major component of senile plaque is a hydrophobic 39–43 amino acid peptide termed the β-amyloid peptide (Aβ), proteolytically derived from a larger membrane-spanning glycoprotein, β-amyloid precursor protein (βPP).

It has been suggested that aberrant processing of βPP and a conformational transition from α-helix to β-sheet in the amino-terminal region of the amyloid β-peptide is linked to the formation of amyloid plaques [155, 156].

4.4 Lysosomal Storage Diseases

Several lysosomal storage diseases result from defective expression of specific glycosidases. The intracellular accumulation of unprocessed carbohydrate substrates leads to aberrant cellular structure and cell death. Thus, deficient or defective hexosaminidase results in either Tay-Sachs or Sandhoff disease [157]. Fucosidosis, another neurovisceral storage disease, is caused by defective α-L-fucosidase expression.

Aspartylglycosaminuria (AGU) is the most common disorder of glycoprotein degradation and is caused by a deficiency of lysosomal glycosylasparaginase, which hydrolyzes β-*N*-glycosidic bond between Asn and Gn, resulting in the accumulation of asparatylglucosamine. The most prominent clinical findings include severe mental and motor retardation. Most cases of AGU are caused by a mutation in the gene that results in a failure to activate the enzyme precursor by intramolecular autoproteolysis [158].

4.5 Other Diseases

A genetic defect of α-M II causes congenital dyserythropoietic anemia type II. The reduction of α-M II activity results in a failure of polylactosaminoglycan formation in erythrocyte membrane proteins, leading to clustering of membrane proteins and formation of unstable erythrocytes [159].

Acute phase proteins are plasma proteins produced mainly by hepatocytes. Most APPs are glycoproteins with one or more *N*-linked complex glycans. Stimulants that commonly induce the acute phase response include tissue injury, rheumatoid arthritis, bacterial infection, inflammation, and neoplasms. Cytokines, notably interleukin-6, induce striking alterations in the concentration and glycosylation pattern of APPs in response to these stimuli [160].

5 Choice of Expression Systems

5.1 Systems Available for Glycoprotein Expression

A number of recombinant proteins including peptide hormones, cytokines, growth factors and monoclonal antibodies have been obtained using different expression vectors and host cell systems [161]. However, it has been observed that the yield and the authenticity of glycosylation vary with the product and the expression system. The choice of expression system for the production of glycoproteins is governed by the following factors:

(a) structural complexity (number of disulfide bonds, oligomerization, extent and type of glycosylation, etc);
(b) product stability and refolding;
(c) desirability of secretion;
(d) intended use (therapeutic or diagnostic);
(e) yield.

In a strict sense, recombinant proteins should be comparable to their counterparts purified from natural sources in terms of biological, clinical and pharmacological properties. As the natures of glycosylation modulate these attributes of a glycoprotein, and no single eukaryotic host cell system is capable of processing every potential heterologous glycoprotein glycans similar to its ectopic tissues, it is imperative to examine various expression systems to find the one that can produce an authentic product. An excellent review on the major glycosylation attributes of different expression systems and the role of culture conditions on glycosylation has recently been published [162].

The acceptable limit of glycosylation differences between natural and recombinant proteins would depend on the type of product, the intended use and the pharmacological attainment. For example, glycosylation in the Fc region rarely affects the immunoreactivity of monoclonal antibodies (MAbs). However, this does affect Fc receptor binding, antibody mediated cytotoxicity and is involved in antibody elimination. While glycosylation of Fab regions has variable effects on the binding activity of MAbs, glycosylation in the hinge region might effect antibody sensitivity to proteases [163].

The most commonly used expression systems are insect cells with baculovirus vectors, yeast and mammalian cells. Filamentous fungi (e.g. *Aspergillus* sp.) have also been developed as host cells for the production of heterologous proteins, although there is need to improve the culture with respect to decrease fungal proteases and glycosylation patterns [164]. Unlike higher eukaryotes, a number of late modification steps, such as Man core processing and addition of Fuc, Gal, Gn and SA are absent in prokaryotes and simple eukaryotes [165].

Expression of recombinant proteins in the milk of transgenic animals is gaining popularity due to simple and cost effective production. Over the past several years, the feasibility of this approach has been demonstrated by the production, at mg ml^{-1} levels, of pharmaceutically relevant monomeric proteins such as human α1-antitrypsin, human tPA, human protein C and hexameric fibrinogen [166]. However there are limitations of mammary tissue in making the meaningful post-translational modifications, which has been overcome by coexpression of key processing enzymes [167].

5.1.1
Yeasts

A wide range of yeast species, including *Saccharomyces cerevisiae, Pichia pastoris, Hansenula polymorpha, Kluyveromyces lactis, Schizosaccharomyces pombe* and *Yarrowia lipolytica* have been tested for the production of heterologous eukaryotic glycoproteins [168, 169]. Major advantages in yeasts include the relatively low culture costs, high expression levels and presence of a secretory pathway similar to that in higher eukaryotes. Yeasts may successfully use secretion signals from other organisms, and secretion of heterologous proteins into the culture medium simplifies product purification due to low levels of native secretory proteins. All the steps involved in higher eukaryotic protein trafficking and post-translational modification may not be equivalent in yeast. Thus, although a number of recombinant glycoproteins with pharmaceutical or industrial value has been obtained using the yeast expression system, they often have altered biological properties and functions with respect to solubility, sensitivity to proteases or serum half-life, mostly due to differences in protein glycosylation. In *S. cerevisiae, N*-linked glycans are processed to highly mannosylated structures and not to complex-type structures containing Fuc, Gal and SA as found in higher eukaryotes [170]. The Man_8Gn_2 oligosaccharide structure is elongated in the Golgi by the action of a series of mannosyltransferases to form the large mannan oligosaccharides, usually containing 50–150 mannose residue and some of them have Man-P-Man sequences [171]. Such glycoproteins are recognized by Man receptors and removed from circulation. In addition, nonhuman glycosylation in expressed proteins are potentially immunoreactive. A mannosylation defective mutant strain has been used to avoid hyper mannosylation. These mutants however, do not grow well like other yeast strains [172]. Species like *P. pastoris* and *H. polymorpha* express glycoproteins of average Man chain length less than 30 monomers, and furthermore the expressed proteins are devoid of highly antigenic terminal α1,3-Man linkage, which is obtained in *S. cerevisiae* system [168]. Besides glycosylation variants, proteolytic instability of the secretory proteins is another significant problem in yeast expression systems. There is a need to develop protease-deficient mutants to improve the quality and yields of the expressed proteins.

5.1.2
Mammalian Cells

The major advantage of using mammalian cells to produce heterologous eukaryotic proteins are that the expressed proteins are correctly folded and glycosylated to near native structures [173]. Because of these unique post-translational modifications, various mammalian expression systems have been reported time to time for the production of therapeutic proteins [173–176]. However, the high cost of cell culture, low productivity (5–50 pg $cell^{-1}$ day^{-1}), and difficulty in growing in large scale imposes serious restrictions on their exploitation. The use of human cell lines for expressing protein is rare. However, the human lymphoblastoid Namalwa cell line has been used for the production of tPA. The cell line performs O-linked and N-linked glycosylation efficiently, and preliminary study shows that the glycosylation pattern is identical to the normal human cell type [177].

5.1.2.1
Mouse Cells

Most mammals express the enzyme α1,3-GalT, which generates Gal α1,3-Gal β1,4-Gn residues on membrane and secreted glycoproteins. Humans are notable exceptions where the gene has become inactivated [178]. Certain mouse cell lines such as hybridomas, mouse-human heterohybridomas, and C127 cells synthesize some glycans terminating in Gal α1,3-Gal β1,4-Gn [179]. Moreover, the levels of N-glycolylneuraminic acid (NeuGc), a SA derivative, are more prevalent in antibodies derived from mouse or human-mouse hybridomas [180]. In contrast, glycoprotein in adult humans do not normally contain NeuGc. Low levels of NeuGc (1% of total sialic acid) are tolerated in recombinant proteins such as EPO, but higher levels (around 7%) can elicit an anti-NeuGc antibody [181]. Furthermore, high levels of terminal NeuGc are correlated with a rapid removal of molecule from circulation, compared to the same protein bearing terminal SA residues [182]. Other rodent cell lines, such as mouse NSO or rat YO myeloma, producing humanized antibodies, do not add this unwanted residue (Gal α1,3-Gal β1,4-Gn) and therefore are only mildly immunogenic [183]. In spite of nonnative forms of glycosylation, mouse cell lines seem to be one of the powerful expression system commercially exploitable. Mouse C127 cell line under BPV promoter has been found to produce tPA at peak titre of 55 mg l^{-1} [184]. The highest yield of antibody (1 g l^{-1}) so far reported is in NSO cells using glutamine synthetase amplification system [185].

5.1.2.2
Hamster Cells

Chinese hamster ovary (CHO) and baby hamster kidney (BHK) cells have a "set" of glycosylation enzymes that are similar (but not identical) to those in human cells [165]. However, these lack a functional α2,6-ST required to add SA through α2,6 linkages. As the rat gene coding for this enzyme has been cloned, the defect can be corrected by α2,6-ST transfection, generating both α2,3- and α2,6-linked SAs [186]. In addition, mutants of CHO cells that display altered glycosylation pattern have proven useful in expressing glycoproteins with minimal heterogeneity [187]. Thus, CHO-K1 cell produces terminally sialylated glycans, while CHO-Lec2 cell produces carbohydrate with a >90% decrease in SA content, and CHO-Lec1 cell, lacks GnT I activity, leading to the accumulation of high-Man intermediates [188]. The highest expression obtained in CHO-K1 cell line was for tPA, and the maximum product concentration obtained was 136.4 mg l^{-1} in a high cell density perfusion culture [189].

5.1.3 *Insect Cells*

The production of heterologous proteins in lepidopteran insect cells using baculovirus expression vector has several highly desirable attributes. The process development time for primary production of a recombinant protein is short, high yield is possible (>30 mg l^{-1}) and the system has the ability to carry out a wide array of post-translational processing [190].

Insect cells lack the capacity to process carbohydrate moieties to "complex" type containing Gal and terminal SA residues, express highly mannosylated proteins, and also are reported to attach xylose [191–193]. They do, however, accomplish the most important stage in the overall folding and quality control of glycoproteins involving glucosidase II and glucosyl transferase [194]. There have been some reports indicating the formation of complex glycosylated recombinant proteins in baculovirus infected insect cells [193]. In addition, there has been a report of successful *O*-linked glycosylation on pseudorabies protein (gp50) following its expression in insect cells [195].

Recombinant baculoviruses, especially *Autographa californica* nuclear polyhedrosis virus (AcNPV), are widely used to express heterologous proteins in a eukaryotic processing environment. They are particularly useful for the high-level expression of eukaryotic proteins with fastidious co- or post-translational processing requirements [194]. The virus infects *Autographa californica* and 30 other insect species. The most commonly used cell line is derived from the fall armyworm, *Spodoptera frugiperda*. Examination of *N*-glycosylation processing of an alternative cell line (*Estigmena acrea*) indicated significant differences in the glycosylation pattern. Except for ST, Ea4 cells possess most of the enzymes involved in production of hybrid and complex *N*-glycans [196]. Traditionally, very late viral promoters (driving expression of polyhedrin or p10) have been used in expression vectors. However, post-translational events such as secretion and glycosylation appear to occur more efficiently when early promoters are used [197–199]. Recently, a new type of baculovirus vector has been developed that can express foreign genes immediately after infection under the control of the promoter from the viral immediately early (ie1) gene. These vectors have been used to modify the *N*-glycan processing capabilities of insect cells by directing the expression of a heterologous processing enzyme (β1,4-GalT) during the early phase of infection. The enzyme then functions as part of insect cell machinery and contributes to the production of a foreign glycoprotein synthesized later in infection with more extensively processed *N*-glycans [200]. Interestingly, GPI membrane anchors are efficiently produced in the baculovirus system [201].

5.1.4 *Plants*

Transgenic plants are emerging as an important expression system for foreign genes. Their major attractions are the potential for protein production on an agricultural scale at competitive cost, stable integration of foreign DNA into the plant genome, and ease of storage of transformed lines as seeds under ambient conditions. A number of groups have expressed antibody fragments, single chain molecules, full length antibodies and immunoconjugates employing a variety of toxins with a view to exploit plants as bioreactors for large scale production [202, 203]. A potato plant expressing enterotoxin vaccine is found to elicit antibody response upon oral immunization in mice [204]. The *N*-linked core high-Man glycans have identical structures in plants and other eukaryotes. Complex plant glycans may be quite heterogeneous, but tend to be smaller than mammalian complex glycans, and differ in the terminal sugar residues. For example, a Xyl residue linked β1,2 to the β-linked Man of the glycan core is frequently found in plants, but not in mammals while SA has not been identified in plants. As a result of asialylated form, EPO produced in tobacco cells has no in vivo biological activity, presumably because of its high clearance rate [205]. The difference in glycosylation pattern of antibodies expressed in plants had no effect on antigen binding or specificity. But, for human therapy, the presence of plant specific glycan might increase the immunogenicity of recombinant antibody. Furthermore, the possibility that patients may develop an allergenic response to plant-derived glycan moieties such as core α1,3-linked fucose need to be taken into account [206]. Enzymatic removals of unwanted glycan groups or the use of mutant or recombinant plant strains with altered glycosylation pathways are ways to approach this problem in cases where inappropriate glycosylation affects the pharmacokinetics, immunogenicity or product efficacy [165, 207].

5.1.5
Dictyostelium discoideum

This well known amoeboid organism is used to express heterologous proteins that are difficult to study in other systems [208]. Various cell lines with different glycosylation capacities have been developed. *D. discoideum* has been successfully used to express two parasite proteins, such as malaria circumsporozoite surface antigen (CSP) [209] and glutathione-*S*-transferase (GST) from *Schistosoma japonicum* [210]. Besides, Rotavirus outer capsid glycoprotein VP7, human glycoproteins antithrombin III and muscarinic receptor have also been expressed. *O*-Linked peptides that are formed in *D. discoideum* are glycosylated at the same residues in humans, though the sugars added are not always the same. This shows a separation of recognition and catalytic domains for the glycosylating enzymes. These studies have assisted in the designation of specific peptide motifs as potential *O*-glycosylation sites [211].

5.2
Protein Glycosylation – Nature of Protein and Host Cells

Each eukaryotic host cell has its own characteristic set of glycosylating enzymes, and thus authentic glycosylation is unlikely to occur in any expression system.

Glycan processing pathways have been genetically manipulated in hosts like *S. cerevisiae* and CHO cells [188]. In general, the natures of glycan structure in the expressed glycoprotein are protein and host specific. The terminal sugars and, to a lesser extent, the chain branching are the features that are most influenced by the host, as observed from the study of tPA and EPO [212]. Table 3a shows the nature of glycans produced in three different glycoproteins expressed in CHO cells. It is obvious that these recombinant proteins differ from their native counterparts with respect to biological activities and circulatory half-lives, mainly due to conformational differences and the nature of glycosylation. The extent to which the host cells dictate the glycosylation differences is illustrated in Table 3b. Therefore, it is important to study the host cells and their glycosylation machinery before considering them for the expression of the therapeutic proteins. This view is further substantiated by the observations with pituitary hTSH, which is having complex glycans terminating predominantly with SO_4-GalNAc. CHO cells do not express GalNAc-transferase and sulfotransferase, and thus hTSH expressed in these contain only SA-Gal terminal sequences and have longer plasma half-life and higher in vivo activity despite lower in vitro activity compared to pituitary hTSH [217].

5.3 Factors Controlling Glycosylation in Cultured Cells

Once the host cell line has been selected, the cell culture conditions are optimized to minimize glycoprotein heterogeneity and to prevent deterioration of product quality [218]. Cell status, bioreactor configuration, and culture conditions (e.g. pH, concentration of NH_4^+) and media components (e.g. serum levels, glucose, amino acids), presence of nucleotide sugars cytidine and uridine, and lipids such as dolichol alone or in combination with lipoprotein carriers, have been found to affect glycosylation in several systems [162]. Large scale cultured tPA exhibited a higher clearance than tPA produced on a small scale, indicating variability in glycosylation pattern [219]. Product degradation due to various glycosidase activities has been measured in CHO cell lysates and culture supernatants. Amongst these, sialidase is the most active enzyme at neutral pH [220] and is found to degrade glycans from recombinant products [221]. Mouse cell lines such as NSO myelomas and hybridomas display much lower sialidase activity than CHO cell at neutral pH.

6 Biopharmaceutical Properties

Recombinant glycoproteins often have altered biological properties and functions, increased immunogenicity and protease sensitivity compared to their native counterparts. Cell lines that produce proteins containing glycans with the blood group antigens or with terminal Gn residues are unsuitable as hosts because of the presence of natural antibodies against blood group antigens and

Table 3 a. Nature of glycosylation in different proteins expressed in the cell line CHO

	Protein expressed	Nature of glycosylation Recombinant protein	Native protein	References
CHO	1. IFNγ	Biantennary complex type (Asn^{25} and Asn^{97})	Two glycosylation sites	[213]
	2. tPA	Oligomannose type chain in one of the three *N*-glycosylation sites. Other two sites (Asn^{184} and Asn^{448}) are complex type	Complex and oligomannose. Three Asn sites, two classes of varients: type I and type II with either Asn^{117} and Asn^{448} or Asn^{117}, Asn^{184} and Asn^{448} respectively. Asn^{117} is oligomannose type, other two are charged complex type	[208] [212, 214]
	3. EPO	Tetraantennary complex type at three Asn sites and one *O*-linked glycans. More *O*-linked glycans and less sialic acid than native form manifest conformational differences and hence in the potency	One *O*-linked and three *N*-linked glycans. *N*-linked glycans are extensively branched and complex type structure having tetra-antennary glycans	[212, 215]

Table 3 b. Nature of glycosylation in IFNγ expressed in different cell lines

Protein espressed	Cell line	Nature of glycosylation	Remarks	References
IFNγ	1. CHO	Complex biantennary type (Asn^{25} and Asn^{97})	The glycans of native IFNγ is exclusively complex biantennary type. Specific activity is comparable with the recombinant proteins, however such heterogeneity in glycans could affect the circulatory half-life due to varying degree of susceptibility to clearance by Man receptors	[191, 195, 215]
	2. Sf9	Mainly trimannosyl core structure		
	3. Mammary gland of transgenic mice	Intermediate glycan profile (complex structure at Asn^{25} and oligomannose structure at Asn^{97})		

the strong antigenicity of Gn terminated glycans. As the glycans with Xyl and Fuc groups, commonly found in plant glycoproteins, are highly immunogenic, recombinant glycoproteins produced by plant cells may also cause pathological concerns [222]. Artificial glycosylation sites have been introduced into small peptides to improve their pharmacokinetic properties or to make them resistant to proteases [223,224]. Although an authentic human glycosylation profile could be the final aim; it may be desirable to produce a modified glycoprotein with defined glycosylation and predictable pharmacokinetic properties [192]. There are four properties critical to the efficacy of therapeutic proteins depending on glycosylation – biological activity, antigenicity, immunogenicity and circulatory half-life. Since the roles of glycosylation in overall biological activities of glycoproteins have been discussed in the earlier sections, they are not separately included here.

6.1 Antigenicity

Correct post-translational processing has a large impact on the efficacy and antigenicity of recombinant protein [225]. Frequently, the α-linked sugars either alone or in combination constitute antigenic determinants. Oligosaccharide structures can serve as a basis for antibody recognition [226]. Glycoproteins with nonnative carbohydrate structures may be antigenic. Many mammalians circulating antibodies are targeted against specific oligosaccharide determinants. Most humans have circulating antibodies against *N*-linked yeast mannan chains, and about 1% of circulating human IgG is specific for the terminal Gal α1,3-Gal epitope generated by mouse C127 cells. There may be rapid clearance due to antigen-antibody complex formation followed by phagocytosis. Glycans may also contribute indirectly to glycoprotein antigenicity. The interaction with antibodies is influenced by their bulky hydrophilic and often highly charged glycan moieties. The glycans appear to adopt a flattened configuration over the surface of the peptide moiety which would tend to maximize steric hindrance in binding to antibody [227]. Glycoproteins such as human α1-acid glycoprotein, which contains 5 glycan moieties essentially of the tri- and tetra-antennary types, are completely enveloped by the glycans. This could explain the resistance and the weak antigenicity of glycoproteins, the glycans acting as protective shields. The umbrella configuration of glycans is firmly maintained by ionic bonds between the electronegative charges of SA residues and electropositive ones of basic amino acids. Removal of SA residues makes the antennae free and mobile, abolishes the protective effect of glycans, and the glycoprotein becomes more antigenic and more susceptible to degradation. The protective role played by glycans towards the protein moieties could explain the resistance of metastatic cancer cells, as their membrane glycoproteins are significantly enriched in tri- and tetra-antennary glycans [228].

6.2
Immunogenicity

The effects of glycans on glycoprotein immunogenicity (ability to elicit an immune response) are less clear [229]. Available evidence shows that the epitopes of glycoproteins consist solely of peptide elements, though the bulky, hydrophilic and often highly charged glycans may influence the conformation and reactivity of glycoproteins. The epitopes are retained in the deglycosylated glycoproteins [230], although there is some evidence that the carbohydrate moiety may influence the immunogenicity of glycoprotein [227]. In a recent study on immunogenicity, it has been observed that nonglycosylated hCGβ elicited better antibody response in rats than that of glycosylated hCGβ [231]. Proteins that are normally glycosylated can potentially have enhanced immunogenicity when administered in aglycosyl form due to a tendency to aggregate [232]. Proteins without appropriate carbohydrate moieties may have altered immunogenicity. Furthermore, absence of glycosylation can unmask peptide epitopes causing an antibody response, as seen with GM-CSF derived from yeast but not that from CHO cells [233].

6.3
Metabolic Clearance and Circulatory Half-Life

Glycans play a significant role in defining the in vivo glycoprotein clearance rate through both specific (receptor mediated) and non-specific (physicochemical) routes, a property critical in determining the efficacy of an injected therapeutic protein. The clearance from circulation is particularly dependent on the presence of glycans on the outer arms [234]. Glycosylation of protein is believed to change specific biological activity, alter diffusibility and tissue distribution and guarantees an effective transfer to target organs. Differences in glycan structure, degree of sialylation and number of antennae play a major role in the clearance rate of an injected glycoprotein and cause differential effects on in vivo activity [216, 234].

Carbohydrate-specific hepatic receptor-mediated clearance mechanisms include the asialoglycoprotein, SO_4-GalNAc and Man receptors [235, 236]. The asialoglycoprotein receptor on liver hepatocytes is most significant and accounts for the short circulatory half-life of proteins lacking terminal SA [237]. Reliable quantitative analysis of SA will be an important aspect of quality control for glycoprotein pharmaceuticals. For glycoproteins produced in recombinant *S. cerevisiae* and insect cells which possess terminal Man or Gn moieties, the Man receptor presumably represents a major clearance mechanism [229].

The glycoprotein hormones hLH, hTSH and the uncombined α-subunit synthesized in pituitary have unusual Asn-glycans with terminal SO_4-GalNAc residues whereas the terminal residues in hFSH and in hCG synthesized in placenta are SA and Gal. A specific GalNAcT in combination with a sulfotransferase accounts for synthesis of Asn-glycans terminating with GalNAc-4-SO_4

[238]. The sulfated glycans found on LH do not affect its bioactivity at the receptor level but do have a marked impact on its circulatory half-life and in vivo potency. A receptor present on hepatic endothelial cells recognizes glycans terminating with SO_4-GalNAc and can account for rapid removal of glycoproteins bearing these structures [239].

In hTSH, *N*-glycans of β-subunit has a more pronounced role than that from α-subunit in metabolic clearance and in vivo activity [240]. Similarly, in β-subunit of hCG, highly sialylated glycans are important in prolonging plasma half-life [241]. The carboxy terminal peptide (CTP) of βhCG is also important in prolonging plasma half-life of the hormone. This has been ligated to hFSH [242] and hTSH [243] to generate analogs with increased plasma half-life and bioactivity. Furthermore, in hTSH, there was a significant decrease in circulatory half-life upon deletion of glycans at $Asn^{\alpha 78}$ compared to at $Asn^{\alpha 52}$ [244], which may be related to surface-exposed location of $Asn^{\alpha 78}$ [134]. These results are comparable with the observed site-specific role of glycans in the in vivo activity of hFSH [245]. The circulatory half-life, and not the in vitro activity, appears to be the primary determinant of the in vivo activity of these hormones.

Glycoprotein glycans can also affect clearance rate by mechanisms which do not involve high-affinity receptors. They can prolong glycoprotein circulatory half-life by increasing both size and surface charge and affecting filtration rate through the kidney glomerular tubules [229]. Furthermore, highly branched glycans (tri- or tetraantennary) are less susceptible to renal clearance than biantennary structures as in EPO [246].

7 Controlled Carbohydrate Remodeling

There is interest in the development of methods that will permit modification of glycan structures on therapeutic glycoproteins. This might increase serum half-life and solubility of drug, decrease antigenicity, promote uptake by target cells and tissues, improve efficacy and reduce dosage [247]. The glycosyltransferases as well as glycosidases have been exploited for the synthesis of oligosaccharides and glycoconjugates. The in vivo function of glycosidases is to cleave glycosidic bonds, although under appropriate conditions they can be useful as synthetic catalysts. Glycosyltransferases are highly specific in the formation of glycosides though, the availability is limited. Glycosidases have the advantage of wider availability and lower cost, but they are not as specific or high yielding in synthetic reactions.

Despite no PCR equivalent replication system being available for the amplification of minute amounts of carbohydrates and no machine being available for the solid phase synthesis of glycans, advances have been made in their chemical synthesis. For example, synthesis of core *N*-glycan structure (Man_3Gn_2) and oligomannose glycans from monosaccharides have recently been accomplished [248–250]. At present it is possible to synthesize carbohydrate chains 20 mers long, to derivatize glycans and link them together, and to employ enzymatic and semisynthetic methods for the generation of lead molecules [251]. This opens the possibility of adding defined glycan structures after recombinant protein synthesis and secretion. In addition, cloning and expression of several important glycosyltransferases in *E. coli* will allow post-harvest remodeling of glycoproteins produced in cell culture, making them more acceptable as human therapeutic proteins [252].

8
Characterization of Sugars

Assessment and control of product heterogeneity are the major problems faced in the manufacture of recombinant proteins. Unraveling the complexities of glycosylation of a molecule is a substantial task. Sequence analysis has a somewhat more entangled meaning for a variably linked and multiple branched structure than for the linear biopolymer as in the case of peptides. Deglycosylation reactions may be discriminating or incomplete and chromatography could enrich or exclude particular subforms. However, in many situations this is not necessary and only a limited amount of information on a single or a group of structural features is needed. Techniques and instruments are now becoming available which, particularly when used in combination, can provide rapid and accurate comparisons of the glycosylation patterns of glycoproteins.

Glycan analysis is performed on intact glycopeptide or after their release from the protein. Thus, two approaches generally practiced to compare the glycosylation pattern of glycoproteins are (a) glycan analysis on intact glycopeptide – site-specific digestion of glycoprotein with proteases or cynogen bromide to yield smaller glycopeptides, separating them, and analysing amino acid sequence and glycan structure by different standard techniques, and (b) glycan analysis – releasing the individual glycans from the glycopeptide or intact protein by chemical or enzymatic treatment, labeling and separating them, and determining the relative quantity of individual released saccharide. Details of procedural aspects for the analysis of glycans are beyond the scope of this review, although a brief description of alternative methods for the analysis and their scope are presented in Table 4. Excellent reviews pertaining to the structural analysis of glycans have been published recently [228, 268].

9
Conclusions

This review has addressed the glycosylation of proteins, its role in biological functions, associated diseases due to defects in protein glycosylation, expression systems available for the production of recombinant glycoproteins and their characterization. Although much progress has been made towards understanding the biological roles of glycosylation and consequences of altered co-/post-translational cellular events, the areas that need to be attended to are the molecular basis of altered glycosylation, and control of glycosylation in the branch

Table 4. Summary of available techniques and their applicability for the analysis of oligosaccharides

Analysis	Techniques followed	Remarks	Ref.
Analysis after digesting protein:			
A) Site specific digestion	1. Treatment with pronase	Yields glycoamino acids or glycopeptides with a very short polypeptide chain	[253]
	2. Treatment with trypsin or Chymotrypsin	Long chain polypeptides are liberated	
	3. Treatment with CNBr	Cleaves at methionine residues	
B) Separation of glycopeptides	1. Gel filtration using BioGel P4 2. Ion exchange chromatography 3. RP-HPLC 4. Lectin affinity chromatography 5. SDS-PAGE gels		
Analysis without digesting protein			
A) Release of glycans	*Chemical release*		
	1. Hydrazinolysis	Applicable for both *N*- and *O*-linked glycans. Peptide bonds are cleaved and amino acids are converted to hydrazides. There is simultaneous cleavage of Gn-Asn linkage and release of acyl groups linked to the amino sugar. Degradation of terminal GalNAc is a major problems with *O*-glycans	[254]
	2. Alkaline β-elimination	Release of only *O*-glycans under mild conditions, and both *N*- and *O*-linked glycans under stronger alkaline conditions	[255]
	3. Hydrogen fluoride	Splits all *O*-glycosyl bonds to yield corresponding monosaccharide fluorides	[256]
	4. Methanolysis	Cleaves all *O*-glycosidic linkages in one-step procedure leading to the formation of *O*-methyl glycosides	[257]

Table 4 (continued)

Analysis	Techniques followed	Remarks	Ref.
	Enzymatic release (N-glycans)		
	1. PNGase F	Cleaves β-aspartylglycosylamine bond of all classes of glycans	[258, 259]
	2. PNGase A	Cleaves sugars having α1,3-linked core fucose, resistant to PNGase F	
	3. Endo H	Selectively cleaves oligomannose and hybride type structure between the two Gn residues within the core	[259, 260]
	4. Endo D	Cleaves all classes of *N*-glycans between the two Gn residues within the core	[259]
	5. Endo F_1, F_2, F_3	Cleave within the chitobiose core, but Endo F_1 digests only oligomannose and hybrid structures, Endo F_2 digests oligomannose and biantennary structure and Endo F_3 is specific for bi- and triantennary glycans.	[259]
B) Labelling of glycans at Reducing termini	*Radioactive labelling*	Classical method, sensitivity of detection increased to femtomole.	
	Tritium	Largely replaced by fluorescent compound labelling	
	Fluorescently labelled	Sensitivity of detection increased and ease in separation due to introduced charge	
	1. 2-aminobenzamide (2-AB)	HPLC, reversed phase (RP)-HPLC, weak anion exchange, high pH anion exchange, BioGel P4 gel permeation chromatography, matrix assisted laser desorption ionisation (MALDI) mass spectrometry (MS) and electron spray (ES) MS could be deployed to separate glycans	[261]
	2. 2-anthranilic acid (2-AA)	Electrophoretic separation of glycans by SDS-PAGE	[262]
	3. 8-aminonapthalene-1,3,6-trisulphonic acid (ANTS)	Particularly useful for electrophoretic separation of neutral sugars, allowing resolution from acidic glycans	[262]
	4. 2-aminopyridine (2-AP)	Used extensively in RP-HPLC and High performance capillary electrophoresis (CE)	[263]
	5. 2-aminoacridone (AMAC)	Suitable for SDS-PAGE of glycans and micellar electrokinetic capillary chromatography (MECC). Also compatible with RP-HPLC and MALDI time-of-flight (TOF) or nanolitre flow ES-MS	[264]

Table 4 (continued)

Analysis	Techniques followed	Remarks	Ref.
C) Separation of glycans	*Chromatography*		
	1. Normal phase HPLC	Capable of resolving subpicomolar quantities of mixtures of fluorescently labelled neutral and acidic *N*- and *O*-glycans simultaneously	[265]
	2. RP-HPLC	The effect of linkage position is much more marked on RP-separations than on normal phase. Oligosaccharides containing bisecting Gn can be identified. Derivatization (e.g. 2-AP for fluorescence detection) to increase hydrophobicity required	[266]
	3. Weak anion exchange HPLC	Suitable for resolving mixture of glycans (above pH 12) on the basis of the number of charged residues. High pH AEC coupled with pulsed amperometric detector (PAD) is routinely used for separating and profiling of sialylated *N*- and *O*-glycans	[267]
	4. Gel permeation (BioGel P4)	Applicable for neutral oligosaccharides. So sample desialylation is essential before use. It allows sequencing of glycans, if associated with the treatment of exoglycosidases	
	5. Lectin affinity chromatography	Used in techniques such as immobilized lectin affinity chromatography	[268]
	Capillary Electrophoresis	Used to separate glycopeptides and released glycan chains with UV- or laser-induced fluorescence (LIF) detection. On of the few techniques able to resolve glycoforms. Suitable as a rapid fingerprinting technique for assessing the glycosylation varients. Oligomannose, hybrid and complex sialylated oligo-saccharides can be resolved	[269, 270]
Structural analysis (applicable to the alternate approaches):			
	1. Digestion of glycopeptides/ glycans with highly specific exoglycosidases	Digestion of each oligosaccharide with enzymes, either sequentially or using enzyme arrays. Extremely sensitive and provides information on all parameters	[271]

Table 4 (continued)

Analysis	Techniques followed	Remarks	Ref.
		[e.g. monosaccharide residue, anomericity (α/β) or linkage, absolute stereoisomer(D/L) of monosaccharide] except substitution pattern. Mono-saccharides are analyzed by normal phase HPLC or MALDI-MS	
	2. NMR spectroscopy	Potentially the most informative for glycan structure, linkage position and anomeric configuration analysis. Relatively insensitive	[272]
	3. Gas chromatography (GC)/MS	Composition and linkage of monosaccharides are best determined after methanolysis	[257]
	4. Fast atom bombardment(FAB) or liquid secondary ion(LSI) MS	Sequence, branching and more linkage information can be obtained. Moderate sensitivity. Sample derivatization necessary	[273]
	5. Tandem MS (MS-MS)	Sequence, branching pattern and some linkage information can be obtained. *O*-linked glycosylation also analyzed	[274]
	6. Electronspray ionization(ESI) MS	Useful in quality control for recombinant proteins and site analysis of tryptic glycopeptides. Readily interfaced with HPLC or CE systems. It can differentiate between *O*- and *N*-linked glycans, and also between complex, hybrid or high mannose forms	[275]
	7. MALDI TOF-MS	Simple, quick and quite insensitive to salts or detergents. Useful for detailed structural analysis including linkage and arm specificity. Unlabelled glycans can be analyzed. Derivatization increases detection limits.	[276]

and terminal regions. Oligosaccharides of glycoproteins are far more complex than the peptide chain. With regard to proteins, DNA serves as the template for RNA synthesis which in turn serves as a template for protein synthesis. In contrast, the biosynthesis of oligosaccharides does not occur from a template or blueprint but results from a complex series of reactions capable of generating many diverse structures. The observation that a single pituitary cell type is capable of synthesizing two distinct glycoprotein hormones which have, at least in part, distinctly different oligosaccharide structures indeed suggests that the processing of oligosaccharides is not a random event.

The glycan structures of recombinant glycoproteins differ with the host cell type, despite having the same polypeptide structure. This is mainly as a result of species-specific and tissue-specific glycosylation. Since the nature of glycosylation could affect the biological activities of proteins, it is preferable to use cell lines for the expression of recombinant therapeutic glycoproteins of identical properties to those of native proteins. There is a need to develop mutants of yeast and insect cells which on one hand do not produce hyper-mannosylated products, and on the other are capable of terminal sialylation of the expressed proteins. In this respect, co-expression of ST and GalNAcT with the protein of interest is important. Concurrently it is also important to exploit and develop mouse NSO cells for expression of glycoproteins. Biological half-life is the most important pharmacokinetic property of a glycoprotein, increasing that will not only reduce the clinical doses of therapeutic proteins, but will also reduce the chances of autoimmune types of diseases. Potentially, by changing amino acids at a maximum of three sites, it is possible to introduce a new glycosylation site and hence to improve available bioactivity of a therapeutic glycoprotein. Attempt have been made to make fusion protein with 37 amino acids from *C*-terminus of hCGβ to improve the biological half-life and hence the available bioactivities of other glycoproteins.

Controlled carbohydrate modeling is a recent exciting development in the area of glycobiotechnology, but at present is too far from the point of utilization. In vitro applications of GlycTs and glucosidases to modify oligosaccharide structures of the purified recombinant glycoprotein is another option for redesigning glycan structure, although it still needs more work before it is technologically and economically feasible.

Acknowledgments. The authors are grateful to the Department of Biotechnology, Government of India for funding the National Institute of Immunology to carry out the work. The authors are also grateful to Professor Abdesh Surolia, Molecular Biophysics Unit, Indian Institute of Sciences, Bangalore and Dr. Chitra Mandal, Indian Institute of Chemical Biology, Calcutta for providing constructive suggestions while this article was being written.

10
References

1. Krishna RG, Wold F (1993) Adv Enzymol-Relat Areas Mol Biol 67:265
2. Wyss DF, Wagner G (1996) Curr Opin Biotechnol 7:409
3. Varki A (1993) Glycobiol 3:97
4. Montreuil J, Vliegenthart JFG, Schachter H (1995) Glycoproteins. Elsevier Science BV, Amsterdam, The Netherlands
5. Allen HJ, Kisailus EC (eds) (1992) Glycoconjugates: composition, structure and function. Marcel Dekker
6. Rothman R, Orci L (1992) Nature 355:409
7. Helenius (1994) Mol Biol Cell 5:253
8. Ruddon RW, Bedows E (1997) J Biol Chem 272:3125
9. Rapoport TA (1992) Science 258:931
10. Walter P, Lingppa VR (1986) Ann Rev Cell Biol 2:499
11. Rapoport TA, Jungnickel B, Kutay U (1996) Annu Rev Biochem 65:271
12. Wickner W (1995) Proc Natl Acad Sci 92:9433
13. Lauring B, Kreibich G, Wiedmann M (1995) Proc Natl Acad Sci 92:9435
14. Corsi AK, Schekman R (1996) J Biol Chem 271:30,299
15. Schatz G, Dobberstein B (1996) Science 271:1519
16. Laboissiere MCA, Sturley SL, Raines RT (1995) J Biol Chem 270:28,006
17. Galat A (1993) Eur J Biochem 216:689
18. Hartl FU (1996) Nature 381:571
19. Todd MJ, Lorimer GH, Thirumalai D (1996) Proc Natl Acad Sci 93:4030
20. Komar AA, Kommer A, Krasheninnikov IA, Spirin AS, (1997) J Biol Chem 272:10,646
21. Chen W, Helenius J, Braakman I, Helenius A (1995) Proc Natl Acad Sci 92:6229
22. Pelham HRB (1995) Curr Opin Cell Biol 7:530
23. Yang Mei, Ellenberg J, Bonifacino JS, AM Weissman (1997) J Biol Chem 272:1970
24. Burke J, Pettitt JM, Humphris D, Gleeson PA (1994) J Biol Chem 269:12,049
25. Ma J, Colley KJ (1996) J Biol Chem 271:7758
26. Cooper MS, Cornell-Bell AH, Chernjavsky A, Dani JW, Smith SJ (1990) Cell 61:135
27. Halban PA, Irminger J-C (1994) Biochem J 299:1
28. Lis H, Sharon N (1993) Eur J Biochem 218:1
29. Hansen JE, Lund O, Engelbrecht J, Bohr H, Nielsen JO, Hansen J-ES, Brunak S (1995) Biochem J 308:801
30. Abeijon C, Hirschberg CB (1992) Trends Biochem Sci 17:32
31. Behrens NH, Tabora E (1978) Dolichol intermediates in the glycosylation of proteins. In: Ginsburg V (ed) Methods enzymol, vol L, part C. Academic Press, NY, p 402
32. Cacan R, Verbert A (1995) Glycosyltransferases of the phosphodolichol pathways. In: Verbert A (ed) Methods on glycoconjugates. Harwood Academic Publishers, Switzerland, p 191
33. Nilsson IM, von Heijne G (1993) J Biol Chem 268:5798
34. Kasturi L, Eshleman JR, Wunner WH, Shakin-Eshleman SH (1995) J Biol Chem 270:14,756
35. Avanov AY (1991) Mol Biol 25:293
36. Gavel Y, von Heijne G (1990) Prot Eng 3:433
37. Shakin-Eshleman SH, Spitalnik SL, Kasturi L (1996) J Biol Chem 271:6363
38. Landolt-Marticorena C, Reithmeier RAF (1994) Biochem J 302:253
39. Allen S, Naim HY, Bulleid NJ (1995) J Biol Chem 270:4797
40. Kornfeld R, Kornfeld S (1985) Assembly of asparagine-linked oligosaccharides. In: Richardson CC, Boyer PD, Dawid IB, Meister A (eds) Annual review of biochemistry, vol 54. Annual Review Inc., Palo Atto, p 631
41. Hammond C, Helenius A (1995) Curr Opin Cell Biol 7:523
42. Beggah AT, Jannin P, Geering K (1997) J Biol Chem 272:10,318
43. Bergeron JJM, Brenner MB, Thomas DY, Williams DB (1994) Trends Biochem Sci 19:124

44. Fiedler K, Simons K (1995) Cell 81:309
45. Herbert DN, Foellmer B, Helenius A (1995) Cell 81:425
46. Hiller MK, Finger A, Schweiger M, Wolf DH (1996) Science 273:1725
47. Bischoff J, Moremen K, Lodish HF (1990) J Biol Chem 265:17,110
48. Velasco A, Hendricks L, Moremen K, Tulsiani DRP, Touster O, Farquhar MG (1993) J Cell Biol 122:39
49. Gerken TA, Owens CL, Pasumarthy M (1997) J Biol Chem 272:9709
50. Kreisel W, Hanski C, Tran-Thi TA, Katz N, Deckers K, Reutter W, Gero KW (1988) J Biol Chem 263:11,736
51. Dwek RA (1995) Biochem Soc Trans 23:1
52. Natsuka S, Lowe JW (1994) Curr Opin Struc Biol 4:683
53. Paulson JD, Colley KJ (1989) J Biol Chem 264:17,615
54. Hounsell EF, Davies MJ, Renouf DV (1996) Glycoconj J 13:19
55. Chou K-C, Zhang C-T, Kezdy FJ, Poorman RA (1995) Proteins: Struct Function and Genet 21:118
56. O'Connell BC, Hagen FK, Tabak LA (1992) J Biol Chem 267:25,010
57. Furuhashi M, Suzuki S, Tomoda Y, Suganuma N (1995) Endocrinol 136:2270
58. Wang Y, Abernethy J, Eckhardt A, Hill R (1992) J Biol Chem 267:12,709
59. Englund PT (1993) Annu Rev Biochem 62:121
60. Udenfriend S, Kodukula K (1995) Annu Rev Biochem 64:563
61. Dwek RA (1996) Chem Rev 96:683
62. Opdenakker G, Rudd PM, Ponting CP, Dwek RA (1993) FASEB J 7:1330
63. Rudd PM, Joao HC, Coghill E, Fiten P, Saunders MR, Opdenakker G, Dwek RA (1994) Biochem 33:17
64. Kern G, Kern D, Jaenicke R, Seckler R (1993) Protein Sci 2:1862
65. Morell AG, Gregoriadis G, Scheinberg IH, Hickman J, Ashwell G (1971) J Biol Chem 246:1461
66. Kelm S, Schauer R, Crocker PR (1996) Glycoconj J 13:1
67. Davis JT, Hirani S, Bartlett C, Reid BR (1994) J Biol Chem 269:3331
68. Otvos L, Thurin J, Kollat E, Urge L, Mantsch HM, Hollosi M (1991) Int J Peptide Protein Res 38:476
69. Imperiali B, Rickert KW (1995) Proc Natl Acad Sci 92:97
70. Doms RW, Lamb RA, Rose JK, Helenius A (1993) Virology 193:545
71. Jaenicke R (1995) Phil Trans R Soc Lond -Biol Sci 348:97
72. Busca R, Pujana MA, Pognonec P, Anwerx J, Deeb SS, Reina M, Vilan S (1995) J Lipid Res 36:939
73. Sadhukhan R, Sen I (1996) J Biol Chem 271:6429
74. Rasmussen JR (1992) Curr Opin Struct Biol 2:682
75. Mathieu ME, Grigera PR, Helenius A, Wagner RR (1996) Biochem 35:4084
76. Li Y, Luo L, Rasool N, Kang CY (1993) J Virol 67:584
77. Billian A (1996) Adv Immunol 62:61
78. Sareneva T, Pirhonen J, Cantell K, Julkunen I (1995) Biochem J 308:9
79. Matzuk MM, Boime I (1988) J Biol Chem 263:17,106
80. Huth JR, Perini F, Lockridge O, Bedows E, Ruddon RW (1993) J Biol Chem 268:16,472
81. Feng W, Matzuk MM, Mountjoy K, Bedows E, Ruddon RW, Boime I (1995) J Biol Chem 270:11,851
82. Blithe DL, Iles RK (1995) Endocrinol 136:903
83. Gowda DC, Jackson CM, Kurzban GP, Mc Phie P, Davidson EA (1996) Biochem 35:5833
84. Hollosi M, Perczel A, Fasman G (1990) Biopolymers 29:1549
85. Avvakumov GV, Warmels- Rodenhiser S, Hammond GL (1993) J Biol Chem 268:862
86. Shogren R, Gerken TA, Jentoft N (1989) Biochem 28:5525
87. Wang C, Eufemi M, Turano C, Giartosio A (1996) Biochem 35:7299
88. Olsen O, Thomsen KK (1991) J Gen Microbiol 137:579
89. Imberty A, Perez S (1995) Protein Eng 8:699
90. Rutherford TJ, Partridge J, Weller CT, Homans SW (1993) Biochem 32:12,715

91. Tahirov TH, Lu T-H, Liaw Y-C, Chen Y-L, Lin J-Y (1995) J Mol Biol 250:354
92. Hecht HJ, Kalisz HM, Hendle J, Schmid RD, Schomburg D (1993) J Mol Biol 229:153
93. Woods RJ, Edge CJ, Dwek RA (1994) Nat Struct Biol 1:499
94. Wyss DF, Choi JS, Wagner G (1995) Biochem 34:1622
95. Davis SJ, Davies EA, Barclay AN, Daenke S, Bodian DL, Jones EY, Stuart DI, Butters TD, Dwek RA, Van der Merwe PA (1995) J Biol Chem 270:369
96. Mer G, Hietter H, Lefevre JF (1996) Nature Struc Biol 3:45
97. Dwek RA (1995) Science 269:1234
98. Malhotra R, Wormald MR, Rudd PM, Fischer PB, Dwek RA, Sim RB (1995) Nature Med 1:237
99. Davis D, Liu X, Segaloff DL (1995) Mol Endocrinol 9:159
100. Rens-Domiano S, Reisine T (1991) J Biol Chem 266:20,094
101. El Battari A, Forget P, Fouchier F, Pic P (1991) Biochem J 278:527
102. Kaushal S, Ridge KD, Khorana HG (1994) Proc Natl Acad Sci 91:4024
103. Recny MA, Luther MA, Knoppers MH, Neidhardt EA, Khandekar SS, Concino MF, Schimke PA, Francis MA, Moebius U, Reinhold BB (1992) J Biol Chem 267:22,428
104. Ding DX-H, Vera JC, Heaney ML, Golde DW (1995) J Biol Chem 270:24,580
105. Fan G, Goldsmith PK, Collins R, Dunn CK, Krapcho MJ, Rogers KV, Spiegel AM (1997) Endrocrinol 138:1916
106. Bradbury FA, Kawate N, Foster CM, Menon KMJ (1997) J Biol Chem 272:5921
107. Hayes GR, Williams A, Costello CE, Enns CA, Lucas JJ (1995) Glycobiol 5:227
108. Davis DP, Rozell TG, Liu X, Segaloff DL (1997) Mol Endocrinol 11:550
109. Servant G, Dudley DT, Escher E, Guillemette G (1996) Biochem J 313:297
110. Kornfeld S (1992) Annu Rev Biochem 61:307
111. Balch WE, McCaffery JM, Plunter H, Farquhar MG (1994) Cell 76:841
112. Sultan A, Miyoshi E, Ihara Y, Nishikawa A, Tsukada Y, Taniguchi N (1997) J Biol Chem 273:2866
113. Kuehn MJ, Heuser J, Normark S, Hultgren SJ (1992) Nature 356:252
114. Schachner M (1989) Ciba Found Symp 145:156
115. Lasky LA (1993) Curr Biol 3:680
116. Anderson A E, et al (1988) Adv Exp Med Biol 228:601
117. Muramatsu T (1988) J Cell Biochem 36:1
118. Devaux P, Loveland B, Christiansen D, Milland J, Gerlier D (1996) J Gen Virol 77:1477
119. Zimmer G, Klenk HD, Herrler G (1995) J Biol Chem 28:17,815
120. Tavakkol A, Burness ATH (1990) Biochem 29:10,684
121. Whelan J (1996) Trends Biochem 65
122. Lasky LA (1995) Annu Rev Biochem 64:113
123. Cremer H, Lange R, Christoph A, Plomann M, Vopper G, Roes J, Brown R, Baldwin S, Kraemer P, Scheff S, Barthels D, Rajewsky K, Wille W (1994) Nature 367:455
124. Opdenakker G, Rudd PM, Wormald M, Dwek RA, van Damme J (1995) FASEB J 9:453
125. Thotakura NR, Blithe DL (1995) Glycobiol 5:3
126. Szkudlinski MW, Grossmann M, Weintraub BD (1996) Trends Endocrinol Metab 7:277
127. Magner JA (1990) Endocrinol Rev 11:345
128. Matzuk MM, Keene JL, Boime I (1989) J Biol Chem 264:2409
129. Bishop LA, Nguyen TV, Schofield PR (1995) Endocrinol 136:2635
130. Grossmann M, Szkudlinski MW, Tropea JE, Bishop LA, Thotakura NR, Schofield PR, Weintraub BD (1995) J Biol Chem 270:29,378
131. Fares FA, Gruener N, Kraiem Z (1996) Endocrinol 137:555
132. Moyle WR, Campbell RK, Venkateswara Rao SN, Ayad NG, Bernard MP, Han Y, et al. (1995) J Biol Chem 270:20,020
133. Jiang X, Dreano M, Buckler DR, Cheng S, Ythier A, Wu H et al. (1995) Structure 3:1341
134. Lapthorn AJ, Harris DC, Littlejohn A, Lustbader JW, Canfield RE, Machin KJ, Morgan FJ, Isaacs NW (1994) Nature 369:455
135. Hoermann R, Keutmann HT, Amir SM (1991) Endocrinol 128:1129
136. Dowbenko D, Andalibi A, Young PE, Lusis AJ, Lasky LA (1993) J Biol Chem 268:4525

137. Sawada T, HoJ J, Chung YS, Sowa M, Kim YS (1994) Int J Cancer 57:901
138. Litscher ES, Wassarman PM (1996) Biochem 35:3980
139. Kozarsky K, Kingsley D, Krieger M (1988) Proc Natl Acad Sci 85:4335
140. Oh eda M, Hasegawa M, Hattori K, Kuboniwa H, Kojima T, Orita T, Tomonou K, Yamazaki T, Ochi N (1990) J Biol Chem 265:11,432
141. Lubas WA, Frank DW, Krause M, Hanover JA (1997) J Biol Chem 272:9316
142. Kreppel LK, Blomberg MA, Hart GW (1997) J Biol Chem 272:9308
143. Troy FA (1992) Glycobiol 2:5
144. Lopez SJJ, Senra VA (1995) Int J Biol Markers 10:174
145. Meyer U, Dierig C, Katopodis N, De-Bruijn CH (1993) Anticancer Res 13:1889
146. Sears P, Wong C-H (1996) Proc Natl Acad Sci 93:12,086
147. Whelan J (1996) Trends Biochem 21:65
148. Youings A, Chang S-C, Dwek RA, Scragg IG (1996) Biochem J 314:621
149. Jaeken J, Carchon H (1993) J Inher Metab Dis 16:813
150. Marquardt T, Ullrich K, Zimmer P, Hasilik A, Deufel T, Harms E (1995) Eur J Cell Biol 66:268
151. Powell LD, Paneerselvam K, Vij R, Diaz S, Manzi A, Buist N, Freeze H, Varki A (1994) J Clin Invest 94:1901
152. Thomas PJ, Qu B-H, Pedersen PL (1995) Trends Biochem Sci 20:456
153. Jilling T, Kirk KL (1997) Int Rev Cytol 172:193
154. Kelly JW (1996) Curr Opin Struc Biol 6:11
155. Saito F, Tani A, Miyatake T, Yanagisawa K (1995) Biochem Biophys Res Commun 210:703
156. Soto C, Castano EM, Frangione B, Inestrosa NC (1995) J Biol Chem 270:3063
157. Sandhoff K, Conzelmann E, Neufeld EF, Kaback MM, Suzuki K (1988) The G_{M2} gangliosidoses. In: Scriver CR, Beaudet AL, Sly WS, Valle D (eds) The metabolic base of inherited diseases, 6th edn. McGraw Hill, NY p 1807
158. Guan C, Cui T, Rao V, Liao W, Benner J, Lin C-L, Comb D (1996) J Biol Chem 271:1732
159. Fukuda MN, Masri KA, Dell A, Luzzatto L, Moremen KW (1990) Proc Natl Acad Sci 87:7443
160. Mackiewicz A (1997) Int Rev Cytol 170:225
161. Koths K (1995) Curr Opin Biotechnol 6:681
162. Jenkins N, Parekh RB, James DC (1996) Nature Biotechnol 14:975
163. Miele L (1997) Trends Biotechnol 15:45
164. Mackenzie DA, Jeenes DJ, Belshaw NJ, Archer DB (1993) J Gen Microbiol 139:2295
165. Jenkins N, Curling EM (1994) Enzyme Microb Technol 16:354
166. Prunkard D, Cottingham I, Garner I, Bruce S, Dalrymple M, Lasser G, Bishop P, Foster D (1996) Nature Biotech 14:867
167. Velander WH, Lubon H, Drohan WN (1997) Sci Amer 276:70
168. Romanos M (1995) Curr Opin Biotechnol 6:527
169. Faber KN, Harder W, Ab G, Veenhuis M (1995) Yeast 11:1331
170. Herscovics AO, Orlean P (1993) FASEB J 7:540
171. Tsai PK, Frevert J, Ballon CE (1984) J Biol Chem 259:3805
172. Lehle L, Eiden A, Lehnert K, Haselbeck A, Kopetzki E (1995) FEBS Lett 370:41
173. Zettmdssl CJ (1987) Bio/Technol 5:720
174. Stephenne J (1989) Production in yeast and mammalian cells of the first rDNA human vaccine against hepatitis B. A technical and immunological comparison. In: Spier RE, Griffiths JB, Stephenne J, Crooy PJ (eds) Advanced animal cell biology and technology for bioprocesses. Butterworths, UK, p 526
175. Mizuguchi M, Matsumoto K, Onodera K (1993) Production of the hGH in serum free UC203 medium. In: Kaminogawa S, Ametani A, Hachimura S (eds) Animal cell technology: basic and applied, vol 5. Kluwer Academic Publishers, Netherlands, p 57
176. Mukhopadhyay A, Mukhopadhyay SN, Talwar GP (1995) Biotechnol Bioeng 48:158
177. Khan MW, Musgrave SC, Jenkins N (1995) Biochem Soc Trans 23:S99
178. Larsen RD, Rivera-Marrero CA, Ernst LK, Cummings RD, Lowe JB (1990) J Biol Chem 265:7055

179. Borrebaeck CAK, Malmborg AC, Ohlin M (1993) Immunol Today 14:477
180. Monica TJ, Williams SB, Goochee CF, Maiorella BL (1995) Glycobiol 5:175
181. Noguchi A, Mukuria CJ, Suzuki E, Naiki M (1995) J Biochem 117:59
182. Flesher AR, Marzowski J, Wang DC, Raff HV (1995) Biotechnol Bioeng 46:399
183. Lifely MR, Hale C, Boyce S, Keen MJ, Phillips J (1995) Glycobiol 5:813
184. Rhodes M, Birch J (1988) Bio/Technol 6:518
185. Werner RG (1994) Successful products and future business prospects. In: Spier RE, Griffiths JB, Berthold W (eds) Animal cell technology: products of todays and prospects for tomorrow. Butterworth-Heinemman, Oxford, p 573
186. Minch SL, Kallio PT, Bailey JE (1995) Biotechnol Prog 11:348
187. Stanley P, Ioffe E (1995) FASEB J 9:1436
188. Stanley P (1992) Glycobiol 2:99
189. Kopp K, Noe W, Schluter M, Walz F, Werner R (1994) Analysis of product consistency: independent of process parameters, tPA shows a stable glycosylation pattern. In: Spier RE, Griffiths JB, Berthold W (eds) Animal cell technology – products of today, prospects for tomorrow. Butterworth-Heinemann UK, p 661
190. O'Reilly DR, Miller LK, Luckow VA (1992) Baculovirus expression vectors: a laboratory manual. WH Freeman, New York
191. Maerz L, Altmann F, Staudacher E, Kubelka V (1995) Protein glycosylation in insect cells. In: Montreuil J, Vliegenhart JFG, Schachter H (eds) Glycoproteins. Elsevier Science BV, Amsterdam, The Netherlands, p 543
192. James DC, Freedman RB, Hoare M, Ogonah OW, Rooney BC, Larionov OA, Dobrovolsky VN, Lagutin OV, Jenkins N (1995) Bio/Tech 13:592
193. Jarvis DL, Finn EE (1995) Virology 212:500
194. Davies AH (1995) Curr Opin Biotech 6:543
195. Thomsen DR, Post LE, Elhammer AR (1991) J Cell Biochem 43:67
196. Ogonah OW, Freedman RB, Jenkins N, Patel K, Rooney BC (1996) Bio/Technol 14: 197
197. Sridhar P, Hasnain SE (1993) Gene 131:261
198. Bonning BC, Roelvink PW, Vlak JM, Possee RD, Hammock BD (1994) J Gen Virol 75:1551
199. Chazenbalk GD, Rapoport B (1995) J Biol Chem 270:1543
200. Jarvis DL, Finn EE (1996) Nature Biotech 14:1288
201. Davies A, Morgan BP (1993) Biochem J 295:889
202. Ma JKC, Hein MB (1995) Trends Biotech 13:522
203. Ma JKC, Hiatt A, Hein M, Vine ND, Wang F, Stabila P, Van Dolleweerd C, Mostov K, Lehner T (1995) Science 268:716
204. Haq TA, Mason HS, Clements JD, Arntzen CJ (1995) Science 268:714
205. Matsumoto S, Ikura K, Ueda M, Sasaki R (1995) Plant Mol Biol 27:1163
206. Altmann F, Tretter V, Kubelka V, Staudacher E, Marz L, Becker WM (1993) Glycoconjugate J 10:301
207. Von Scaeven A, Sturm A, O'Neill J, Chrispeels MJ (1993) Plant Physiol 102:1109
208. Williams KL, Emslie KR, Slade MB (1995) Curr Opin Biotech 6:538
209. Reymond CD, Beghdadi-Rais C, Roggero M, Duarte EA, Desponds C, Bernard M, Groux D, Matile H, Bron C, Corradin G, Fasel N (1995) J Biol Chem 270:12,941
210. Dittrich W, Williams KL, Slade MB (1994) Biotechnol 12:614
211. Gooley AA, Williams KL (1994) Glycobiol 4:413
212. Geisow MJ (1992) Trends Biotechnol 10:333
213. Warren CE (1993) Curr Opin Biotechnol 4:596
214. Guzetta AW, Bara LJ, Hancook WS, Keyt BA, Bennett WF (1993) Anal Chem 65:2953
215. Kung CKH, Goldwasser E (1997) Protein SFG 28:94
216. Drickamer K (1991) Cell 67:1029
217. Szkudlinski MW, Thotakura NR, Bucci I, Joshi LR, Tsai A, East-Palmer J, Shiloach J, Weintraub BD (1993) Endocrinol 133:1490
218. Jenkins N (1996) Curr Opin Biotechnol 7:205
219. Mueller HS, Rao AK, Forman SA (1987) J Am Coll Cardiol 10:479

220. Gramer MJ, Goochee CF (1993) Biotechnol Prog 9:366
221. Warner TG, Chang J, Ferrari J, Harris R, Mcnerney T, Bennett G, Burnier J, Sliwkowski MB (1993) Glycobiol 3:455
222. Furukawa K, Kobata A (1992) Cur Opin Biotechnol 3:554
223. Takei Y, Chiba T, Wada K, Hayashi H, Yamada M, Kuwashima J (1995) J Interferon & Cytokine Res 15:713
224. Baudys M, Uchio T, Hovgaard L, Zhu EF, Avramoglou T, Jozefowicz M Rihora B, Park JY, Lee HK, Kim SW (1995) J Control Release 36:151
225. Luckow V (1991) In: Prokop A, Bajpai RK, Ho CS (eds) Recombinant DNA technology and applications. McGraw Hill, p 97
226. Feizi T, Childs RA (1987) Biochem J 245:1
227. Storring PL (1992) Trends Biotechnol 10:427
228. Verbert A (ed) (1995) Methods on glycoconjugates. Harwood Academic Publ, Switzerland
229. Goochee CF, Gramer MJ, Andersen DC, Bahr JB, Rasmussen JR (1991) Bio/Technol 9:1347
230. Schwarz S, Krude H, Klieber R, Dirnhofer S, Lottersberger C, Merz WE, Wick G, Berger P (1991) Mol Cell Endocrinol 80:33
231. Mukhopadhyay A, Bhatia PK, Mazumdar SS (1998) Am J Repod Immunol 39:172
232. Jelkmann W (1992) Physiol Rev 72:449
233. Gribben JG, Devereux S, Thomas NS, Keim M, Jones HM, Goldstone AH, Linch DC (1990) Lancet 335:434
234. Flesher AR, Marzowski J, Wang WC, Raff HV (1995) Biotechnol Bioeng 46:399
235. Pontow SE, Kery V, Stahl PD (1992) Int Rev Cytol 137:B221
236. Fiete D, Srivastava V, Hindsgaul O, Baenziger JU (1991) Cell 67:1103
237. Lodish HF (1991) Trends Biochem 16:374
238. Mengeling BJ, Manzella SM, Baenziger JU (1995) Proc Natl Acad Sci 92:502
239. Smith PL, Baenziger JU (1992) Proc Natl Acad Sci 89:329
240. Szkudlinski MW, Thotakura NR, Weintraub BD (1995) Proc Natl Acad Sci 92:9062
241. Hoermann R, Kubota K, Amir SM (1993) Thyroid 3:41
242. Fares FA, Suganuma N, Nishimori K, La Polt PS, Hsueh AJW, Boime I (1992) Proc Natl Acad Sci 89:4304
243. Joshi L, Murata K, Wondisford FE, Szkudlinski MW, Desai R, Weintraub BD (1995) Endocrinol 136:3839
244. Grossmann M, Szkudlinski MW, Tropea JE, Bishop LA, Thotakura NR, Schofield PR, Weintraub BD (1995) J Biol Chem 270:29,378
245. Bishop LA, Nguyen TV, Schofield PR (1995) Endocrinol 136:2635
246. Misaizu T, Matsuki S, Strickland TW, Takeuchi M, Kobata A, Takasaki S (1995) Blood 86:4097
247. Danishefsky SJ, Bilodeau MJ (1996) Angew Chem 35:1380
248. Meldal M (1994) Curr Opin Struc Biol 4:710
249. Matsuo I, Nakahara Y, Ito Y, Nukada T, Ogawa T (1995) Bio-organic and Med Chem 3:1455
250. Nakahara Y, Shibayama S, Ogawa T (1996) Carbohydrate Res 280:67
251. Boons GT (1996) Drug Dev Today 1:331
252. Galili U, Anaraki F (1995) Glycobiol 5:775
253. Rush RS, Derby PL, Smith DM, Merry C, Rogers G, Rohde MF, Katta V (1995) Anal Chem 67:1442
254. Patel TP, Parekh RB (1994) Release of oligosaccharides from proteins by hydrazinolysis. In :Lennarz WJ, Hart GW (eds) Methods enzymol, vol 230. Academic Press, NY, p 57
255. Hermentin P, Doenges R, Witzel R, Hokke DH, Vliegenthart JF, Kamerling JP, Conradt HS, Nimtz M, Brazel D (1994) Anal Biochem 221:29
256. Kakehi K, Ueda M, Suzuki S, Honda S (1993) J Chromatogr 630:141
257. Merkle RK, Poppe I (1993) Carbohydrate composition analysis of glycoconjugates by gas-liquid chromatography/mass spectrometry. In: (ed) Methods enzmol, vol 230. Academic Press, NY, p1
258. Alexander S, Elder JH (1989) In: Ginsburg V (ed) Methods enzymol, vol 179. Academic Press, NY, p 505

259. O'Neill RA (1996) J Chromatogr A 720:201
260. Mellors and Sutherland (1994) Trends Biotechnol 12:15
261. Bigge JC, Patel TP, Bruce JA, Goulding PN, Charles SM, Parekh RB (1995) Anal Biochem 230:229
262. Jackson P (1996) Mol Biotechnol 5:101
263. Honda S, Makino A, Suzuki S, Kakehi K (1990) Anal Biochem 191:228
264. Jackson P (1991) Anal Biochem 196:238
265. Guile GR, Rudd PM, Wing DR, Prime SB, Dwek RA (1996) Anal Biochem 240:210
266. Hicks KB (1988) Adv Carbohydr Chem Biochem 46:17
267. Townsend RR, Basa LJ, Spellman MW (1996) Identification and characterization of glycopeptides in tryptic maps by high-pH anion exchange chromatography. In: Karger BL, Hancock (eds) Methods enzymol, vol 271. Academic Press, NY, p 135
268. Rudd, PM, Dwek RA (1997) Curr Opin Biotechnol 8:488
269. Novotny MV (1996) Glycoconjugate analysis by capillary electrophoresis. In: Karger BL, Hancock WS (eds) Methods enzymol, vol 271. Academic Press, NY, p 319
270. Masada RI, Skop E, Starr CM (1996) Biotechnol Appl Biochem 24:195
271. Edge CJ, Rademacher TW, Wormald MR, Parekh RB, Butters TD, Wing DR, Dwek RA (1992) Proc Natl Acad Sci USA 89:6338
272. Field MC, Amatayakul-Chantler S, Rademacher TW, Rudd PM, Dwek RA (1994) Biochem J 299:261
273. Kovacik V, Hirsch P, Kovac P, Heerma W, Thomas-Oates J, Haverkamp J (1995) J Mass Spectr 30:949
274. Reinhold BB, Haver CR, Plummer TH, Reinhold VN (1995) J Biol Chem 270:13,197
275. Schindler PA, Settineri CA, Collet X, Fielding CJ, Burlingame AL (1995) Protein Sci 4:791
276. Harvey DJ, Naven TJP, Kuster B, Bateman RH, Green MR, Critchley G (1995) Rapid Commun Mass Spectrom 9:1556

Bioaffinity Based Immobilization of Enzymes

M. Saleemuddin

Department of Biochemistry, Faculty of Life Sciences and Interdisciplinary Biotechnology Unit, Aligarh Muslim University, Aligarh – 202 002, India. *E-mail: btisamu@x400.nicgw.nic.in*

Procedures that utilize the affinities of biomolecules and ligands for the immobilization of enzymes are gaining increasing acceptance in the construction of sensitive enzyme-based analytical devices as well as for other applications. The strong affinity of polyclonal/monoclonal antibodies for specific enzymes and those of lectins for glycoenzymes bearing appropriate oligosaccharides have been generally employed for the purpose. Potential of affinity pairs like cellulose-cellulose binding domain bearing enzymes and immobilized metal ion-surface histidine bearing enzymes has also been recognised. The bioaffinity based immobilization procedures usually yield preparations exhibiting high catalytic activity and improved stability against denaturation. Bioaffinity based immobilizations are usually reversible facilitating the reuse of support matrix, orient the enzymes favourably and offer the possibility of enzyme immobilization directly from partially pure enzyme preparations or even cell lysates. Enzyme lacking innate ability to bind to various affinity supports can be made to bind to them by chemically or genetically linking the enzymes with appropriate polypeptides/domains like the cellulose binding domain, protein A, histidine-rich peptides, single chain antibodies, etc.

Keywords: Concanavalin A, Monoclonal Antibodies, Polyclonal Antibodies, Reloadable Biosensors, Antibody Orientation, Fusion Proteins, Glycoenzymes, Immobilized Metal Ion Supports, Enzyme Stabilization.

Advances in Biochemical Engineering / Biotechnology, Vol. 64
Managing Editor: Th. Scheper

List of Abbreviations

BSA	Bovine serum albumin
CBD	Cellulose binding domain
Con A	Concanavalin A
Glc	Glucose
GlcNAc	*N*-acetyl glucosamine
HRP	Horse radish peroxidase
IDA	Iminodiacetic acid
IgG	Immunoglobulin gamma
IMA	Immobilized metal affinity adsorption
Man	Mannose
NTA	Nitricotriacetic acid

1 Introduction

Developments in the areas of recombinant DNA technology, protein engineering and more recently solvent engineering [1, 2] have remarkably enhanced the potential of enzymes in catalyzing the transformation of both water soluble and water insoluble substances, for industrial and analytical applications. Most of the enzyme applications necessitate homogeneous or at least reasonably pure preparations that tend to be expensive. Recovery, reuse and stabilization of the enzymes in the usually unphysiological or even hostile environments to which they are exposed during operation therefore becomes obligatory in order to make the transformations and analyses cost-effective. The now rather mature enzyme immobilization technology offers a spectrum of strategies and a judicious choice amongst these is likely to enhance the performance of any given enzyme [3, 4]. While interest in irreversible and covalent methods of immobilization continues, bioaffinity based procedures are gaining remarkable attention especially for analytical applications [5,6]. Compared to other methods of immobilization, those based on bioaffinity offer several distinct advantages. These include:

1. Binding of the enzyme to affinity support may be very strong yet reversible under specific conditions.
2. Immobilization process is usually simple, mild and necessitates no special skills.

3. Possibility of reuse of the support matrix.
4. Possibility of oriented immobilization facilitating good expression of activity and stabilization against inactivation.
5. Possibility of direct immobilization of the enzyme from partially pure preparations or even crude homogenates.

Selective binding to appropriate ligand is presumably the single common feature of all proteins and enzymes are no exception. While enzymes bind with remarkable specificity and strength to their substrates, cofactors, inhibitors or their analogues, such affinities can not be normally utilized for immobilization as the binding may block the active site of enzyme and thus interfere with catalytic process. Need therefore exists for ligands that bind to epitopes located at a distance from the active site. Table 1 lists the affinities of several biomolecules that have been utilized for enzyme immobilization and whose affinities for enzymes appear to be adequate to keep them immobilized during operation and/or storage. Among these, specific antienzyme antibodies are clearly the most versatile being applicable, at least theoretically, to virtually every enzyme. Affinity of lectins – particularly concanavalin A (Con A) towards several glycoenzymes has also been made use of, for their immobilization. While other biomolecules with affinity towards enzymes have been utilized to a smaller extent the potential of at least some of them appears certainly remarkable.

2 Immunoaffinity Immobilization of Enzymes

Among the numerous applications of biospecific adsorption, immunoadsorption is the most exciting and its potential in enzyme purification has long been recognized. Specific antibodies can be raised against any enzyme in suitable experimental animals and they can be utilized after appropriate screening for the immobilization of the enzyme on suitable support. In view of the availability of mild and simple procedures of protein-protein conjugation, even the affinities of antibodies raised against unrelated proteins can be utilized for immo-

Table 1. Association constants of some bioaffinity pairs

	K_{assoc}	Reference
Antibody -Hapten	10^5-10^{11}	[7]
Antibody -Antigen	10^5-10^{11}	[8]
Lectin-Carbohydrate		
Simple sugars	10^3-10^4	[9]
Macromolecules	10^6-10^7	[9]
Avidin -Biotin	10^{15}	[10]
Protein A -IgG	10^6	[11]
Immobilized metal-Histidine	10^3-10^6	[12]

Adopted principally from [5].

bilization of various enzymes that can be linked either to the antigen or the antibody [13]. Considerable know-how regarding the immunization, isolation of specific antibodies and their immobilization has emerged principally in connection with the immunoaffinity purification of enzymes and this can be readily adopted for immunoaffinity immobilization of various enzymes.

2.1 Monoclonal or Polyclonal Antibodies?

Both monoclonal and polyclonal antibodies have been employed in the immobilization of enzymes and their relative merits and limitations examined (Table 2). Hybridoma cultures can provide a spectrum of monoclonal antibodies from which appropriate population can be conveniently screened and their continuous supply ensured from the selected hybridoma clones [17]. Culturing of hybridomas continues to however remain an expensive endeavour due to the high costs of culture media and experimental set up required. Polyclonal antibodies on the other hand can be relatively inexpensive especially if they can be raised in a large animal like pig or goat. Heterogeneity of the poly-

Table 2. Enzymes immobilized favourably with the help of antibodies

Enzyme	Antibody	Support matrix	Stabilization against	Ref.
β-Galactosidase	PC	–	Temperature	[14]
Gulonolactone oxidase	PC	–	*In vivo* Proteolysis	[15]
Transglutaminase	PC/MC	Agarose beads	Temperature	[16]
Carboxy peptidase A	MC	Eupergit C	Temperature, pH, storage in cold	17, 19]
Lactate dehydrogenase	MC	Eupergit C	Storage	[19]
Glucose oxidase	MC	Sepharose	Recycling	[13, 20]
Chymotrypsin	PC	Sepharose	–	[21]
Subtilisin	PC	–	Sodium hypochloride Temperature	[22]
Nitrate reductase	MC	Sepharose	Temperature	[23]
Alpha amylase	PC	–	Temperature, lyophlization Freezing & Thawing	[22, 24]
Glucoamylase	PC	–	Temperature, Ethanol	[22, 24]
Horse radish peroxidase	MC	–	–	[25]
Trypsin	PC	Sepharose	–	[26]
Urease	PC	Nylon	pH	[27]
NAD glycohydrolase	PC	Nylon	Storage	[27]
Invertase	PC	Sepharose	Temperature	[28]
	Glycosyl Specific PC	Sepharose	Temperature	[29, 30]
L-Hydantoinase	PC	Sepharose	Incubation	[31]

PC -Polyclonal; MC –Monoclonal.

clonal antibody population is however more of a rule than exception, although this can be advantageous in some situations. While methods exist for the fractionation of polyclonal antibodies recognising different epitopes of a single antigen or those of differing in affinity for a single epitopes [32, 33], handling of large volumes of antisera may be problematic especially if the required antibody constitutes only a minor fraction of the total antibody population. For most enzyme immobilization applications, however, heterogeniety of the antibody population may not cause any serious problem provided they do not comprise of inhibitory and/or labilizing antibodies. Formation of active site recognising and hence inhibitory antibodies is quite likely if an animal is immunized with a native enzyme [17, 34, 35], although several reports describing the non-inhibitory nature of the antisera raised against several enzymes are available [14, 24, 28, 36–39]. Some plausible explanations offered for the absence of inhibitory antibodies in the antisera include: the active site acting as blind spot for the immune system, steric hinderance by high affinity antibodies recognising adjacent locations of active site directed antibodies and continued accessibility of the active site in the complex formed between active site recognising antibodies and the enzyme [22].

Some ingenious techniques for the prevention of the formation of active site directed polyclonal antibodies have also been described in the recent years. Fusek et al. [21] immunized pigs with active site blocked chymotrypsin prepared by treating the enzyme with diisopropylphosphofluoridate. The non-inhibitory antibodies were isolated from the sera of immunized animals using chymotrypsin coupled to Sepharose via its active site as the affinity adsorbent. More recently Stovicova et al. [26] raised non inhibitory anti-trypsin antisera in pigs by immunizing them with trypsin complexed with its specific inhibitor antilysin. The IgG fraction isolated from the sera of the immunized animals on coupling to Sepharose support yielded an immunosorbant that immobilized trypsin without decreasing its catalytic activity.

The potential of polyclonal glycoenzyme glycosyl recognizing antibodies in the immobilization of glycoenzymes has also been demonstrated. It was envisaged that since enzyme glycosyls participate rarely if at all in catalytic function [40], glycosyl recognising antibodies may not be inhibitory and hence useful in the immobilization of enzymes. Convenient procedures are available for raising antiglycosyl polyclonal antibodies against glycoproteins and glycoenzymes [29, 41, 42]. Briefly the strategy involves preparation of neoglycoproteins comprising of the oligosaccharides of the enzyme in question and a polypeptide region of a simple protein like BSA. Animals are immunized against the neoglycoproteins and their antisera passed through affinity column of a suitable support to which the enzymes are coupled. Only the antibodies exhibiting affinity towards the enzyme glycosyls are retained in the column and they are eluted using appropriate chaotropic agents or by change in pH [29]. Glycosyl recognising polyclonal anti-invertase antibodies were found to be non-inhibitory towards the enzyme and more effective in the immobilization of the enzyme than those recognising the polypeptide domains [29, 30].

To my knowledge there exists a single report describing the direct comparison of the relative effectiveness of monoclonal and polyclonal antibodies in the immobilization of an enzyme. Ikura et al. [16] observed that activity of guinea pig liver transglutaminase affinity bound on monoclonal antienzyme antibody support was higher than that bound to matrix precoupled with polyclonal antibodies. The authors however have not reported if the polyclonal antibody preparation used contained inhibitory or labilizing antibodies. More detailed comparison is however available in case of non-enzyme proteins- bovine and human serum albumins. After investigating the effects of a variety of factors on the adsorption equilibrium between immobilized polyclonal and monoclonal antibodies and the antigens it was concluded that the former showed a homogeneous affinity of Langmuir type while the later was heterogeneous. It was also observed that binding on immunoadsorbants prepared using polyclonal antibodies may be relatively stronger and their dissociation more difficult due to the binding of the antigen to more than one kind of antibodies [43, 44]. In instances where a single enzyme has the possibility of interacting with more than one molecule of antibody as is the situation during soluble [22] or insoluble immunocomplex formation [15, 28], non inhibitory polyclonal antibodies may offer considerable advantage. In such situations polyclonal antibodies recognising more than one epitope of the enzyme may fix the native conformation of the later and hence confer greater stability against denaturation [22, 28]. It is now well recognized that enzymes attached to support via multiple covalent [45] or non-covalent associations [46] exhibit remarkably higher stability than those attached via fewer linkages. Additive stabilising effects of some monoclonal antibodies have also been demonstrated recently [47]. Monoclonal antibodies may have shorter half lives [48] and as compared to the polyclonals they may be more labile [20].

2.2 Selection of the Antibodies

Animals immunized with enzymes may respond by producing a spectrum of antibodies differing not only in their specificities towards various epitopes but also with respects to their affinities. The dissociation constants of antigen and antibody complexes have been shown to vary between $10^{-3}-10^{-14}$ mol $\cdot$ dm^{-3} [49]. Antibodies of intermediate avidities are considered optimal for preparative affinity chromatography in order to ensure adequate binding specificity and good recovery of the bound enzyme/protein antigen [50, 51] and those in the range of $10^{-6}-10^{-10}$ mol $\cdot$ dm^{-3} are taken as particularly useful [52]. Antibodies with relatively higher affinities may be better suited for the immobilization of enzyme, as in case of immunodiagnostic analysis where essentially irreversible binding is required. Monoclonal antibodies exhibiting affinities falling within the required ranges can be readily isolated by immunoaffinity purification from the chosen hybridoma clones [53, 54]. Reports describing fractionation of the polyclonal antibodies based on their affinity towards the antigenic determinants are also available [33, 55, 56]. Such fractionation is however of only limited value if the desired antibody population does not constitute a major fraction of those present in the antiserum.

Some attempts have also been made to obtain polyclonal antibodies of desired affinity from the sera of immunized animals. For instance, polyclonal antibodies derived from rabbits until 20 weeks of priming were shown to be of low or intermediate affinity compare those isolated after longer durations [57, 58]. Emomoto et al. [59] have more recently demonstrated that binding to the enzyme of antibodies from primary response is far more sensitive to pH and ionic strength alterations than those from the secondary response indicating significant differences in the nature of binding.

2.3
Immunoaffinity Enzyme Immobilization Strategies

A variety of approaches have been adopted for the immobilization of enzymes with the help of antibodies. These range from simple complexing of enzymes with their antibodies to form insoluble complexes to enzyme immobilization on matrices with oriented antibodies.

2.3.1
Immobilization Without Solid Support

Enzyme-antibody complex formation represents the simplest among the immunoaffinity immobilization procedures and the immunocomplexes can be readily formed simply by mixing of the enzyme solution with the antibody or even antiserum. Interestingly neither pure enzyme nor pure antibody may be required for the formation of immunocomplexes. Several early [14, 36, 39] and some recent studies [22, 24, 28] indicate high retention of catalytic activities by various enzymes in the immunocomplexes and marked stability enhancement against various forms of inactivation. The small particle dimensions of the enzyme-antibody complexes may however lead to their compact packing and consequently to slow flow-rates in the column reactors. Their usefulness can be however remarkably enhanced by entrapping the complexes in a polymeric matrix [60, 61].

2.3.2
Construction of Immunoadsorbents

Majority of immunoaffinity enzyme immobilization studies employ specific antibodies coupled to appropriate porous/non-porous solid supports in order to help facilitate ready masstransfer, heat transfer, etc. and offer good flow characteristics. The matrix associated antibody generally acts as a large spacer contributing remarkably to the accessibility of the immobilized enzyme. A repository of useful information on the preparation of immunoaffinity supports is available from the studies aimed at developing immunoaffinity adsorbents for enzyme purification. Since the first deployment of immobilized antibody in enzyme purification [62], a large number of variations and adaptations for the improvement of the strategy have appeared. Excellent reviews are available on the subject [63, 64]. More recently, Ehle and Horn [65] and Desai [52] reviewed comprehensively the work in the area including that on the choice of antibodies, preparation of immunosorbents and optimal conditions for binding and elution of enzymes from the support.

Preparation of immunoaffinity adsorbents require reasonably pure antibodies and usually salt/solvent fractions of antisera [28, 33] or affinity purified polyclonal/monoclonal antibodies [18,22] are employed. Antibody purification strategies have been reviewed extensively [53, 65] and do not fall with in the purview of this article. It is however of interest to point out of a novel strategy employing thermo-sensitive immunomicrospheres that appear highly effective in the large scale purification of antibody from the sera of immunized animals [66, 67].

Majority of the immunoadsorbents constructed for the purpose of enzyme immobilization comprise of appropriate supports to which are attached the antibodies using group specific reagents [23, 26–28, 67]. While such immunoadsorbents serve as efficient supports, random binding of the antibodies on to the support may remarkably lower the ability of the antibody to bind the enzyme. For instance Ikura et al. [16] have shown that the binding capacity of the anti-transglutaminase monoclonal antibody for the enzyme may be lowered to about one fifth on random coupling to Affi-Gel10. Several attempts have therefore been made to favourably orient antibodies on the support matrices.

2.3.2.1
Favourable Orientation of Antibodies on Supports

The oligosaccharide chains of polyclonal antibodies are primarily [68] though not exclusively [69] located in their Fc regions. Several investigations have been therefore directed towards immobilizing antibodies via their carbohydrates chains for improving the accessibility of the antigen binding sites. Although Quash et al. [70] were the first to immobilize IgG through oligosaccharide moieties, Prisyazhnoy et al. [71] made the first comparison between the rabbit anti-mouse antibodies immobilized through the -SH and via oligosaccharides and concluded that the latter exhibited a 3-fold higher antigen binding activity. Similarly, anti-human IgG immobilized on hydrazide derivative via oligosaccharide also exhibited superior binding activity towards the IgG [72]. Little et al. [73] have however observed that the magnitude of increase in antigen binding activity depends upon the nature of antigen-antibody pair. In a more recent study Fleminger et al. [74] immobilized carboxypeptidase and horse radish peroxidase on amino and hydrazide derivative of Eupergit C via carbohydrate chains and achieved antigen binding activity close to the theoretical value of 2 moles antigen bound/mole immobilized antibody. Surprisingly, comparative increase in binding activity was not observed when attempts were made to orient monoclonal antibodies by immobilizing them through their oligosaccharide chains [75, 76]. This may be related to the nature and extent of glycosylation of the monoclonal antibodies [77].

An alternative strategy of oriented immobilization IgG appears to be their binding to the matrices precoupled with protein A that recognizes exclusively the Fc region [78]. Solomon et al. [17] prepared, purified and characterized several mouse monoclonal antibodies recognising carboxypeptidase and compared the properties of two carboxypeptidase preparations immobilized with the help of mouse monoclonal antibody – m100 which was either randomly coupled to the support or oriented favourably on protein A-Sepharose [18].

While no attempts were made to compare the binding capacity of matrices, enzyme immobilized on either support were shown to exhibit full catalytic activity, improvement in stability, and no alteration in K_m, V_{max} and K_i values. The preparation obtained by direct covalent coupling of carboxypeptidase to support however exhibited relatively lower activity. It was also observed that orientation of the antibody on protein A supports may not after all enhance its enzyme binding activity. The author reasoned that even the most accessible antibody molecule may not bind to more than one enzyme molecule due to the large dimensions of the latter.

Favourable orientation of antibodies also appears possible on immobilized metal ions supports. Hale and Beidler [80] observed that an innate histidine-rich sequence located in the C-terminal portion of the Fc region, well-conserved in every antibody classes investigated from several species, binds strongly to the Co^{2+}-IDA resin. Thus, antibodies bound on the supports are oriented with their combining site directed away from the resin and facilitate maximum antigen binding [81].

The need for the minimization of non-specific adsorption and optimization of the binding of the enzymes to antibody support has also been addressed in some studies. DeAlwis et al. [13], and deAlwis and Wilson [82] suggested the use of Fab' fragments instead of the intact antibody. As the immobilization of antibody via its side chain amino groups involves the risk of modifying or blocking the antibody binding sites, coupling of the Fab' fragments through the hinge region thiol groups on supports activated with 2,2,2-trifluoroethanesulfonyl chloride [82] or maleimide [71] was also proposed. At pH 6.0 most amino acid side chain amino groups of protein are protonated and hence unavailable for reaction with tresyl activated supports. Coupling therefore occurs preferentially and predominantly via the thiol groups [82].

2.3.3
Use of Secondary Antibodies

Several investigators have also utilized secondary antibodies in addition to the primary antibodies for immunoaffinity enzyme immobilization. Two strategies were employed by deAlwis et al. [13] for the immunoaffinity immobilization of glucose oxidase in a flow injection system using an immunological reactions. The first of these involved immobilization of human IgG on activated controlled pore glass followed by the binding of enzyme-anti IgG conjugate. Alternatively, a monoclonal antiglucose oxidase antibody immobilized on the support precoupled with Fab' fragment of antimouse IgG was used as an immunoadsorbent for the preparation of immobilized glucose oxidase. The anti IgG-glucose oxidase conjugate used in the first procedure comprised of a mixture of those in which a single enzyme molecule was linked to one, two or even three IgGs [83]. Their binding onto the IgG support was apparently multipoint and gave a more stable immobilized preparation. The second approach on the other hand yielded a far more versatile support capable of binding all IgGs. Both types of bioreactors were effectively used for the measurement of glucose concentration in undiluted sera. deAlwis and Wilson [20] immobilized avidin on Reactigel on which was retained the biotin bound secondary antibody. An immune complex of the enzyme-antiglucose oxidase or antienzyme antibody followed by enzyme was passed through a small column of the Reactigel in order to achieve immobilization. In view of the remarkably high affinity of the avidin-biotin system, antigen-antibody interactions could be conveniently disrupted for enzyme desorption without affecting the matrix-secondary antibody interaction. The glucose oxidase reactors prepared thus, when coupled to a flow injection analysis system could be used for the sensitive and reproducible analysis of glucose.

2.4
Reusability of Immunoaffinity Supports

Relative stabilities of monoclonal and polyclonal antibodies may differ substantially [20, 47], yet their stability against various forms of inactivation are usually far superior than those of the antigenic enzymes. The severity of the elution procedure permissible in the immunoaffinity purification of enzymes is therefore decided by its effect on the later rather than on the antibody. Since desorption of the immunoaffinity immobilized enzyme from the reactors or sensors is undertaken only when it fails in its catalytic function, elution condition far harsher than those employed in affinity purification of enzyme may be permissible. Presumably for this reason, non-specific elution procedures have been generally applied with remarkable success for elution of enzymes immobilized on antibody supports. These include 0.1 M phosphoric acid pH 2.0 [13], 0.2 M glycine buffer pH 2.3 [26], 6.0 M urea in buffered saline, 3.0 M guanidine HCl or 0.1 M acetic acid [16] or 10% dioxane pH 2.5 and 0.5% (v/v) Triton X-, pH 7.0 [27]. Even while employing non-specific elution procedure for elution of enzymes, the immunoaffinity supports have been used successfully for over 50 binding and elution cycles without any decrease of binding capacity [52] and for over 200 cycles with a decrease of about half of the initial binding capacity [84].

A few instances of multiple reuse of immunoaffinity supports of bioreactors/biosensor are also available. deAlwis et al. [13] eluted glucose oxidase from a polyclonal antibody support in a flow injection analysis system at acid pH and reloaded the enzyme for 10 cycles without any apparent loss in binding. Similarly immunoaffinity bound urease and NADase could be eluted and fresh enzyme bound to the reactor for 5 cycles with no decrease in binding [27]. The binding capacity of the anti-transglutaminase antibody support remained unharmed after 4 elution and binding cycles [16]. In the last two studies it was also demonstrated that specific binding of the antigenic enzyme to the appropriate antibody support can be achieved directly and specifically from the crude homogenates. This suggests that pure enzyme preparations may not be essential for the immobilization of enzymes on immunoaffinity support.

Non-specific enzyme elution procedures may however gradually inactivate and decrease the utility of antibody supports. For instance, exposure of antibody supports to pH below 4.0 may decrease the affinity for the antigen and promote susceptibility to proteolysis [52]. Improvement in resistance to proteolysis has however been achieved by controlled modifications of matrix associated antibody with polyethylene glycol [85].

2.5
Behaviour of the Immunoaffinity Immobilized Enzyme

High immobilization yields, expression of high activity of the bound enzyme and stabilization against inactivation/ denaturation are the principal yardsticks for the measurement of the success of any enzyme immobilization procedure and have received considerable attention in case of immunoaffinity immobilized enzymes as well. As has been already pointed out, the antibody molecule acts as large spacer holding the enzyme at a distance from the support matrix thereby minimizing steric hindrance and facilitating remarkable freedom to act even on high molecular weight substrates [18]. Expression of nearly full activity by the enzyme complexed directly with antibody [22, 24] or when bound to support matrix-coupled antibody have been observed by several investigators [26]. In addition, the K_m values of several immunoaffinity immobilized enzymes were either unaltered, exhibited minor alterations as compared those of their respective soluble enzymes. This has been observed in the case of trypsin [26], transglutaminase [16], NADase and urease [27] and carboxypeptidase [17, 18]. The last study also described relatively unaltered K_i values of the enzyme for small molecular weight inhibitors. Table 2 lists the impressive enhancement in the stabilities of the immunoaffinity immobilized enzymes against high temperature, chemical denaturants, proteolysis oxidative stress, etc. Antibody mediated stabilization may not however be a general phenomenon as reports of labilizing antibodies are also available.

Where applicable the stability enhancement may arise out of crosslinking like effect caused by the antibody binding on enzyme. Antibody binding appears to involve reasonably large areas of protein antigen that in turn may comprise more than a single oligopeptide [86, 87]. Based on their studies on the interaction on the influenza virus protein and the Fab' fragment of the antibody, Davis et al. [88] suggested that a large number of non-covalent interactions are involved and water molecules are eliminated from the contact areas. Shami et al. [22] argued that reduction in the free energy of the antigen resulting from the binding of even a moderate affinity antibody [89] may be sufficient to confer stability as free energy changes between the folded and the unfolded states of protein lie in the same range [90]. Furthermore although due to the large size of the antibody a single matrix bound antibody may not bind more than one molecule of enzyme, lateral interactions with more than one antibody molecule may contribute significantly to the stability of the enzyme protein [91]. It is interesting to note that some monoclonal antibodies exhibit chaperon like activity and assist in antigen refolding [92, 93] and inhibit enzyme aggregation [94].

In instances where soluble antibody is used for the enzyme immobilization, a single enzyme molecule may interact with more than one antibody molecules resulting in high degree of stabilization [24] like an enzyme attached via multiple covalent [45] or non-covalent linkages [46]. Indeed the degree of stabilization achieved with such complexes is relatively very high [22, 28]. While a correlation may exist between the thermal stability of protein and its susceptibility to proteolysis [95], the exact mechanism by which antibodies enhance stability against other forms of inactivation needs to be further examined.

2.6 Utility in Organic Solvents

The potential of enzyme action in organic solvents is now well recognized. Some enzymes have been shown not only to retain their catalytic function but also to catalyze novel reactions and exhibit remarkable stability in organic solvents [2]. The significance of support matrices for the utilization of enzymes in organic solvents has been recognized [96] and proteinic supports may be particularly useful [97]. While no detailed study on the behaviour of enzymes immobilized on antibody support in organic solvents has been, to my knowledge, reported, antibodies seem to retain their affinity at least for haptens. Russel et al. [98] demonstrated that the binding and specificity of the hapten 4-aminobiphenyl for the immobilized monoclonal antibody 2E-is retained in several non-aqueous media but the interaction was found to be inversely related to the hydrophobicity of the solvent. Similarly the monoclonal antibody 8 H 11, that binds to the pesticide aldrin, coupled to a large molecular weight carbohydrate support retained its ability bind the pesticide albeit with lowered affinity in organic solvent for 5 h; over 90, 60 and 57% binding was retained in acetonitrile, methanol and 2 propanol respectively [99]. In an earlier interesting study Janda et al. [100] observed that immobilization on porous glass conferred additional stability to lipase like catalytic monoclonal antibody in organic solvents.

3 Lectin Affinity Based Immobilization of Glycoenzymes

Lectins and glycoenzymes represent the second bioaffinity pair with proven potential in the immobilization of the later. Glycosylation is the most common post-translational modification encountered in eucaryotes, archaebacteria and even some procaryotes [40] and a large number of enzymes are glycosylated to varying extent [40]. Several enzymes being currently used in industry and analysis are glycoproteinic [101].

To date hundreds of lectins have been isolated from plants, microorganisms and animals. In spite of some common biological properties that lead to assignment of majority of lectins to distinct families of homologous proteins, lectins and their sugar recognizing/combining sites may be structurally diverse [102]. The largest and best characterized among the lectins is Con A belonging to the legume family. Con A has been widely used in the affinity purification of a variety of glycoenzymes and nearly exclusively in the immobilization of glycoenzymes on a variety of supports. Work on the utility of Con A in glycoenzyme immobilization has been recently reviewed [103].

3.1 Concanavalin A

Con A is a non-glycosylated oligomeric protein that exists as tetramer above pH 7.0 but dissociates to dimers below pH 6.0. Each monomer is 237 amino acids long and has sites for binding of a transition metal ions and a calcium ion in

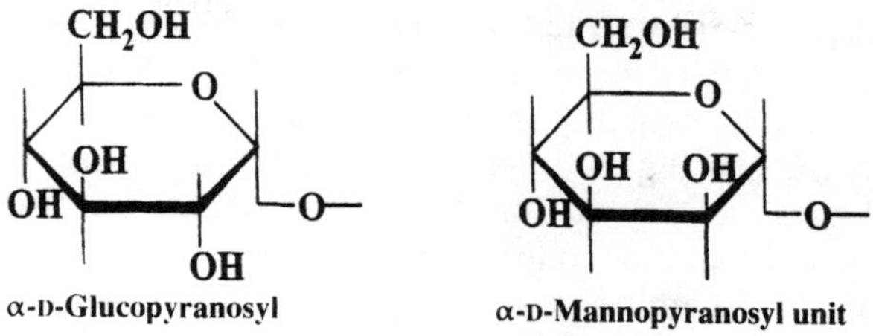

Fig 1. Carbohydrate units recognised by concanavalin A

addition to the saccharide binding site [104]. Early studies on the lectin indicated that Con A recognizes α-D-glucose and α-D.mannose with free hydroxyl at 3-, 4-and 6- positions [105] indicating the structural requirements as in Fig. 1. Glycoproteins comprising of glycosyl chains with Glcα1→, Manα1→ and GlcNAcα1→residues which are located at the non-reducing termini of the sugar chains and →2Glcα1- and →2Manα1-residues located within the sugar chain can bind to this lectin [106]. Based on the binding of several oligosaccharides and glycopeptides, Ogata et al. [107] suggested that high mannose type of sugar chains bind more strongly than those of complex type and presence of at least two binding residues is essential for a oligosaccharide to be retained by a column of immobilized Con A.

Con A retains its affinity for carbohydrates between pH 5.0 or lower and above pH 9.0 [108] making it suitable for the immobilization of enzymes acting over a wide pH range. A recent study has however suggested that inspite of the comparable affinities of the tetrameric/dimeric forms for simple carbohydrate ligands the affinity towards more complex oligosaccharides is considerably higher in case of the tetrameric Con A [109]. This suggested that the Con A supports may be relatively more effective for the immobilization of enzymes acting optimally above pH 7.0.

3.2
Immobilization of Glycoenzymes Using Concanavalin A

The principal strategies of enzyme immobilization with the help of Con A are very similar to those that employ antibodies, i.e. formation of insoluble lectin-glycoenzyme flocculate and binding to insoluble supports precoupled with lectins [103]. Due to the multivalent nature of Con A and the presence of several oligosaccharides on most glycoenzymes, Con A readily interacts with glycoenzymes in solution to form large insoluble lectin-glycoenzyme complexes.

Since enzyme glycosyls, through which the enzymes are associated with lectins are not usually involved in catalytic process [40], the Con A enzyme complexes retain high catalytic activity [110]. Remarkable stability is also exhibited by enzymes complexed with lectins [60] presumably due to the fixation of the enzyme molecules in the native state by several lectin molecules in the complex [103]. The Con A-glycoenzyme complexes however have small particle dimensions and may necessitate additional immobilization procedures [60]. As Con A-glycoenzyme complex formation does not require high purity of either the glycoenzyme or lectin, possibility of cutting down of the cost of immobilization has been demonstrated using crude lectin preparation in place of pure Con A [111].

The biological activity of Con A seems to remain unaffected by chemical modification of the amino groups [112]. Coupling to supports has therefore been achieved mostly via amino groups. A large number of support matrices have been used for the coupling of Con A for the preparation of the affinity supports [103]. More recently Fadda et al. [113] have shown that Con A can be coupled in high yield to commercial aluminium oxide after activation by means of 3-aminopropyltriethoxysilane and subsequent reaction with cyanogen bromide.

Con A has been shown to bind very strongly to copper (II) iminodiacetic acid functions of various supports due to the presence of six histidine residues in each subunit [114]. Such binding has also been shown to favourably orient Con A on the support facilitating very high degree of binding to glycoproteins [115].

A simple strategy for remarkably raising the amount of Con A and glycoenzyme associated with solid supports has been recently developed by us [116]. It involves building up layers of Con A and glycoenzymes on Sepharose matrix precoupled with Con A (Fig. 2). Bioaffinity layered preparation of glucose oxidase and invertase exhibited high activity, good mechanical stability and a layer-by-layer increase in thermal stability. The technique was applicable to all the glycoenzymes investigated namely-glucose oxidase, invertase, β-galactosidase and amyloglucosidase. A somewhat similar approach was earlier used by Gemeiner et al. [117] in the preparation of superactive immobilized invertase.

Enzymes immobilized on Con A supports show impressive gains in resistance to inactivation induced by heat, chemical denaturation, proteolysis, storage [103, 118] and long term continuous operation for several weeks [119]. Taking into consideration the stabilization that accompanies immobilization of glycoenzymes via their carbohydrate chains using lectins [103], glycosyl-specific antibodies [29, 30] or by chemical reaction [120, 121], it appears likely that regions close the glycosylation sites of enzymes may be important in the unfolding of several glycoenzymes. It is now recognized that unfolding of protein molecules may begin at specific region of the molecule and improvement in stabilization may be remarkable if such regions are blocked by attachment to suitable supports [122, 123].

Binding of Con A to glycoproteins can be prevented/reversed by several sugars like methyl-α-D-glucopyranoside, α-D-mannopyranoside or even D-mannose or D-glucose which are known to interact with the lectin [124, 125]. The strength of association between Con A and glycoprotein has however been shown to be dependent not only on the nature of glycosylation but also on the nature of glycoprotein [126]. Some glycoenzymes like invertase exhibit extraordinary strong binding and necessitate extra long incubation with the eluting sugar [127]. It was also shown that ease of elution of the enzyme depends upon the lectin concentration on the support. Nevertheless several lectin-based analytical devices with reloadable glycoenzymes have been described.

Mattiasson and Borreback [128] employed glucose oxidase and peroxidase in a thermistor devices containing immobilized Con A for the measurement of

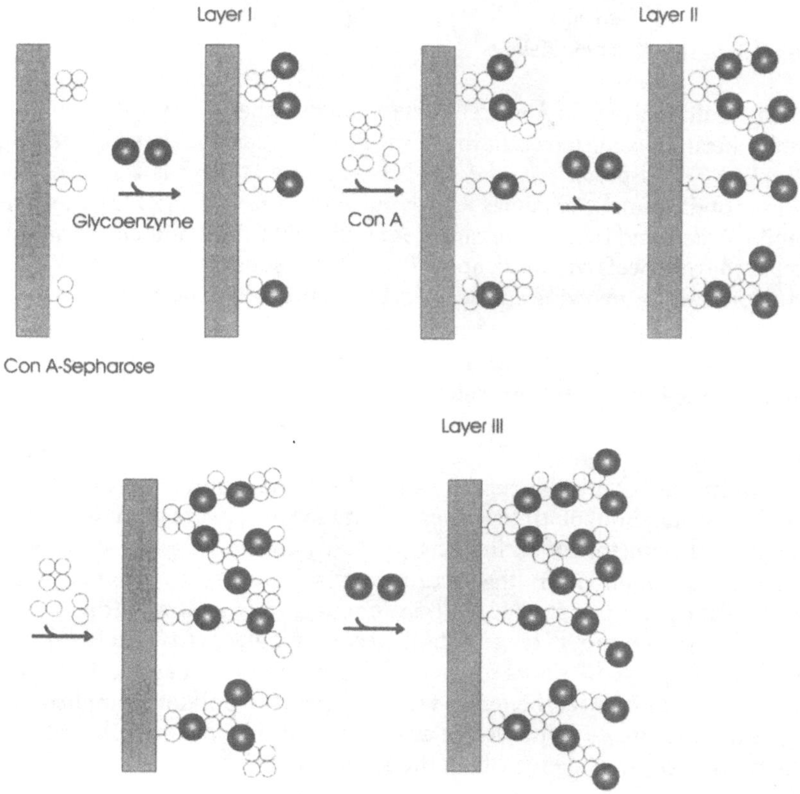

Fig 2. Schematic representation of the bioaffinity layering of glycoenzyme on a concanavalin A support

glucose and peroxide. An effective enzyme thermistor containing Con A Sepharose and the glycoenzyme L-ascorbate oxidase was also described [129]. More recently Köneke et al. [130] described a fluoride sensitive field-effect transistor ($Si/SiO_2/Si_3N_4/LaF_3$ layers) based biosensor with Con A immobilized on the basic membrane for the continuous measurement of glucose concentration. The glycoenzymes glucose oxidase and peroxidase were bound on Con A immobilized on the surface on the sensor. The enzymes could be readily removed from the sensor surface with acetate solution pH 1.0 and the sensor reloaded with fresh enzyme. The biosensor could be successfully integrated in a flow-injection analysis system for continuous monitoring of glucose during the cultivation of *S. cerevisiae.* We have also recently employed a cartridge containing bioaffinity layered glucose oxidase on Con A-Sepharose support connected to an oxygen electrode for the sensitive measurement of glucose concentration during the fed-batch of cultivation of the *S. cerevisiae* on a synthetic medium [116].

3.3 Immobilization of Whole Cells

The high affinity of Con A for cell surface oligosaccharides has also facilitated the immobilization of various cells including those of yeast [131], red blood cells [126, 128] and *Trichosporon cutaneum* [132]. An early study has also described the co-immobilization of enzymes and living cells using Con A [133]. More recently Habibi-Rezaei and Nemat-Gorgani [134] immobilized submitochondrial particles prepared from beef liver mitochondria on Con A support for continuous catalytic transformations involving succinate-cytochrome c reductase.

3.4 Immobilization Using Other Lectins

Very few reports are also available on use of lectins other than Con A in the immobilization of glycoenzymes. *Lens culinaris* lectin that has Con A like sugar specificity but about 50 times lower affinity for the sugars than does Con A was used in a thermistor for facilitating the ready elution of glucose oxidase [128]. The *Cajanus cajan* lectin, that also recognizes glucose and mannose but with lower affinity than Con A [135] on immobilization yields an affinity adsorbant that effectively immobilizes and stabilizes several glycoenzymes [136]. Kokufuta et al. [61] employed *Ricinus communsis* lectin for a novel application alongwith *E. coli* β-galactosidase; the lectin was entrapped in polyion complex-stabilized alginate gel beads to improve the availability of the enzyme substrate *O*-nitrophenyl-β-D-galactoside for which the lectin has affinity.

4 Enzyme Immobilization Using Other Bioaffinity Supports

4.1 Enzyme Immobilization on Immobilized Metal Ion Supports

Immobilized metal affinity adsorption (IMA) is a collective term that was proposed to include all kinds of adsorptions whereby metal atoms or ions immobilized on polymers cause or dominate the interactions at the sorption site [137, 138]. Introduced originally by Porath and coworkers [139] IMA has been widely used in the chromatography/fractionation of proteins and is among the most popular methods available todate [140]. IMA chromatography has also several other applications including in the separation of red blood cells [141] and recombinant *E. coli* cells containing surface hexahistidine clusters [142]. The potential of IMA in enzyme immobilization is also being increasingly realized.

IMA is based upon affinity of surface functional groups of protein for the immobilized metal ions. The strength of association between the chelated and bound metal ion and the residues on protein surface may be quite high and even approach the strength of interaction between enzyme and cofactor/inhibitor or even that between antigen and antibody [140]. Considering that affinity of proteins and enzymes for the chelated metal ions may not be completely non-physiological it has been aptly termed pseudo bioaffinity [12].

First row transition metals (Ni^{2+}, Zn^{2+}, Co^{3+}, Cu^{2+}, Fe^{3+}) are generally used as ligands after chelation by iminodiacetic acid (IDA). IDA is a trident chelator that binds metal ions through its nitrogen and two carboxylate oxygens. As the metal ions coordinate 4–6 ligands, the remaining of coordination sites are occupied by water molecules and buffer ligands which are available for replacement by appropriate surface groups [140].

Although a number of amino acid side chain groups are known to bind to chelated metal ions, it is now recognized that the histidine side chain dominate in protein binding to chelated Ni^{2+}, Zn^{2+}, Co^{2+} and Cu^{2+}. In addition the *N*-terminal amino groups as well as the -SH may contribute to the binding although the latter may not be available in the reduced state on the surface of majority of enzymes. Indeed a number amino acid residues may contribute to metal affinity binding especially at higher pH even though at the commonly used pH (around neutrality), their contribution may be quite small [140].

Considerable evidence has now accumulated to show that protein are retained by a metal affinity supports according to the number of accessible histidine residues [140, 143]. The histidine imidazole nitrogens co-ordinate metals in the unprotonated state hence in addition to the number of accessible histidines, their ionization state, that is in turn dependent upon the nature of surrounding amino acids, influence the retention of protein on metal chelate supports [143].

The association of proteins and enzyme with metal chelate is also strongly influenced by metal ions involved in the coordination. Affinity of the metal for imidazole usually forms the basis of protein retention and stability constants. The affinity of imidazole for Cu^{2+} is about 15 times higher than for Zn^{2+} and Co^{2+} [144, 145]. Although histidine containing proteins bind more strongly to metal chelated supports bearing Cu^{2+}, protein binding capacities of various immobilized metal ion supports may not be very different. In a more recent study, Hale [81] has shown that proteins could be attached to Co^{2+} supports in an exchange inert or irreversible fashion when the cobalt is oxidised from the Co^{2+} to Co^{3+} state. It was however indicated that the protein bound Co^{3+} support may be removed by reduction of the metal to Co^{2+} by appropriate reducing agents.

Elution of proteins bound to immobilized metal supports can be easily performed either with Lewis acids (H^+, Zn^{2+}), that compete with metal for protein or with Lewis bases like imidazole that compete with protein for metal. Elution can also be performed with strong chelators of metal ions like EDTA [140].

Histidine is one of the less abundant amino acids of proteins and most globular proteins contain not more that 2% histidine [146]. Considering that less than half of histidine residues of proteins are located at the surface, most proteins on the average have only about one exposed residue for every hundred amino acids. It is therefore natural that many proteins do not bind to metal chelates. Most of the work on binding and immobilization to metal chelate supports is therefore related to proteins with enriched histidines. However it has been shown that a single surface histidine may be adequate to retain a protein strongly on metal chelate supports [140].

Histidine-rich or polyhistidine affinity tails can be associated with proteins and enzymes by fusing the coding sequence of the former with those of the latter [147, 148]. Such fusion proteins could be immobilized by taking advantage of the specific binding ability of the affinity peptide to appropriate IMA supports.

Ljunquist et al. [149] observed that β-galactosidase fusion protein bearing a histidine-rich peptide extension at *N*-terminus was bound to Zn^{2+}-IDA adsorbent and retained significant enzyme activity. The activity was however lost once the enzyme was eluted, which according to the authors, was the result of inhibition of the enzyme by Zn^{2+} leaking from the IDA supports. In order to overcome the problems of metal ion bleeding, Piesecki et al. [150] immobilized β-galactosidase fusion protein comprising of six histidine residues at the *N*-terminus to Ni^{2+} -nitricotriacetic acid (NTA) adsorbent that binds metal ion far more strongly than the IDA adsorbent [151]. NTA groups form tetracoordinated complexes with metal ions leaving two free coordination sites, while IDA groups form tricoordinate complexes leaving three coordination sites [138]. The association between metal ion and support may thus be stronger in case of NTA substituent, while that between protein and metal loaded support is stronger in case of supports having IDA [81].

Carlsson et al. [152] in a more recent study determined the activities of several enzymes with genetically attached polyhistidine tails bound to metal chelate supports. These included lactate dehydrogenase $(his)_4$, galactose dehydrogenase $(his)_5$, β-glucouronidase $(his)_4$ and the complex prepared using protein A $(his)_5$ and horse radish peroxidase labelled immunoglobulin (protein A $(his)_5$ IgG HRP). All enzymes exhibited good activity in the immobilized state but protein A $(his)_5$ IgG HRP was most active and exhibited essentially same activity as the soluble complex both on Cu^{2+} and Zn^{2+} supports. Activities of galactose dehydrogenase$(his)_5$ and β-glucouronidase $(his)_4$ were high if Zn^{2+} instead of Cu^{2+} was used on the chelating material. Enzymes however showed a lower tendency to bleed from the chelated copper supports. All immobilized enzymes exhibited good storage stability which was at least in the same range as that of the respective soluble enzyme.

A chemical procedure for the enhancement of the surface histidine content of glycoenzymes in order to improve their affinity for metal chelate supports is also available [153]. The method involves oxidation of the carbohydrate moiety of the glycoprotein with periodate followed by covalent modification with histidine. Using the strategy it was possible to increase remarkably the affinity of *Pencillium chrysogenum* glucose oxidase for metal ion chelated supports and confer affinity to horse radish peroxidase that does not bind to the support in the native form. The histidine-enriched enzymes exhibited high activity, remarkable stability and could be used effectively in the analysis of the glucose.

Several proteins and enzymes have innate affinity in their native state for the metal chelate supports due to the presence of one or more surface histidines (Table 3). Coulet et al. [154] immobilized lactate dehydrogenase, malate dehydrogenase and alkaline phosphatase on Co^{2+}, Zn^{2+}, Cu^{2+}-chelate Sepharose. All the enzymes were active in the immobilized state although the retention of the specific activity was quite low. Among the enzymes used, alkaline phosphatase retained higher activity and among the various metal chelates maximum activi-

Table 3. Native and affinity-tail bearing enzymes immobilized on metal chelate supports

Enzyme	Support	Ref.
Lactate dehydrogenase	Cu^{2+}-IDA-agarose	[154]
Malate dehydrogenase	Cu^{+} -IDA-agarose	[154]
Alkaline phosphatase	$Cu^{2+}/Zn^{2+}/Co^{2+}$-IDA-agarose	[154]
Ribonuclease A	Cu^{2+}-IDA-silica	[114]
Lysozyme	Cu^{2+}-IDA-silica	[114]
Glucose oxidase	Cu^{2+}-IDA-silica	[114]
Ribonuclease B	Cu^{2+}-IDA-silica	[114]
Peroxidase $(his)_n$	Cu^{2+}superose HR10/2	[153]
Glucose oxidase $(his)_n$	Cu^{2+}superose HR10/2	[153]
β-galactosidase $(his)_6$	Ni^{2+} NTA adsorbent	[150]
β-galactosidase (Ala-His, Gly-His-Arg-Pro)n	Zn^{2+} NTA adsorbents	[149]
Galactose dehydrogenase $(his)_5$	Cu^{2+}/Zn^{2+} IDA-Sepharose	[152]
Lactose dehydrogenase $(his)_4$	Cu^{2+}/Zn^{2+} IDA-Sepharose	[152]
β-glucouronidase $(his)_4$	Cu^{2+}/Zn^{2+} IDA-Sepharose	[152]

ty was retained on Co^{2+}-chelate agarose followed by that with Zn^{2+}; Cu^{2+} supports were most inferior in this regard.

Anspach and Hasse [155] reported the immobilization of penicillin G amidohydrolase from *E. coli* on immobilised metal chelate supports. A number of the other proteins and enzymes are also known to bind strongly to metal chelate supports including ribonuclease A and lysozyme [114].

Immobilized metal chelate supports have also been used in the preparation of affinity sorbents for enzymes, that do not exhibit high affinity for the immobilized metal ions, by binding to immobilized proteins that strongly bind both to the carriers and to the enzymes. The "sandwich affinity sorbents", so called because the affinity ligands is located between the immobilized metal chelate and the enzyme, have been used principally in the biospecific interaction chromatography of enzymes and proteins [114]. Con A has been shown bind strongly to Cu^{2+} iminodiacetic acid functions in view of the presence of six histidine residues in each subunit [156] and retains affinity towards carbohydrates and glycoproteins. Infact Con A appears to be oriented in a manner that facilitated high accessibility of the carbohydrate binding sites for combining with carbohydrates and glycoproteins [115]. Con A immobilized on metal chelate supports has been shown to bind a number of enzymes including peroxidase and glucose oxidase [114]. Hale et al. [80] demonstrate that antibodies bind to Co^{2+}-IDA-support via metal binding sites located on C- terminal portion of the heavy chain. This facilitated the orientation of the antibody with antigen binding site facing away from the support. It was possible to immobilize antibodies in an exchange inert fashion by oxidising the antibody bound Co^{2+} to Co^{3+} and all the investigated subclasses of the antibodies could be immobilized by this procedure [81]. Considerable potential of enzymes immobilized on immunoaffinity support exist in industry and analysis [157] and the Co^{2+}-chelate supports may be very useful in preparing supports with favourably oriented antibodies for the purpose.

4.2 Immobilization of Chimeric Enzymes

Ability to bind to appropriate ligands can be conferred on enzymes, that lack affinity for the ligand, by linking them to polypeptides that exhibit high affinity for the ligand. Such chimeric enzymes may be affinity bound on supports bearing appropriate ligands. Fusion proteins and enzymes have been widely used in the purification of cloned gene products [158, 159] and their remarkable potential in the enzyme immobilization is clearly evident from some recent investigations.

The exoglucanase (Cex) and endoglucanase (CenA) of *Cellulomonas fimi* bind strongly to microcrystalline cellulose through their cellulose binding domain (CBD)[160]. The CBD functions quite independently of the catalytic domain from which it is separated by a 22-aminoacid long Pro-Thr box [161]. Ong et al. [162] constructed a fusion protein by genetically linking the genes of β-glucosidase of *Agrobacterium sp.* and *Cellulomonas fimi.* The fusion protein retained 42% of β-glucosidase activity, bound strongly to cellulose supports and facilitated efficient substrate hydrolysis. The hybrid enzyme remained stably adsorbed to Avicel for upto 7 weeks at 4 °C and 37 °C, retained activity upto 70 °C and in upto 1.0 M salt solution. The enzyme could be readily eluted with distilled water [163]. More recently Phelps et al. [164] chemically coupled CBD to glucose oxidase with the help of glutaraldehyde and demonstrated the concept and feasibility of a reloadable biosensor based on reversible immobilization of enzyme using CBD technology. While genetic fusion ensures uniformity and homogeneity of the conjugate as well as simplification of its production, size and ratio of the two proteins may be more easy to control in chemical coupling procedures.

A hybrid protein comprising of protein A and β-lactamase has also been constructed using recombinant DNA technology [165]. The hybrid enzyme bound specifically and strongly to IgG-Sepharose and the immobilized preparation was superior in its ability to hydrolyse penicillin G as compared to the covalently immobilized enzyme.

Neusted et al. [166] in an interesting recent study have shown that resistance to trypsin inactivation can be engineered in L-asparaginase through production of chimeric protein between the enzyme and a protective single chain antibody. Although the objective of the investigation was not to immobilize the enzyme the work clearly indicates the potential antibody-enzyme conjugates in the stabilization and immobilization of enzymes.

Acknowledgements. Parts of the author's work cited in the review were supported by the Indian Council for Medical Research, Department of Science and Technology, CSIR India and the Volkswagen Foundation, Germany. The author is grateful to Prof. Dr. T. Scheper of Institute fuer Technische Chemie der Universitat Hannover for many helpful discussions and suggestions.

Note added in proof. It has been shown in a recent study that differences between affinity towards metal ions chelated by Sepharose-linked immunodiacetic acid groups could be made use of to bind two enzymes in a manner that facilated their selective elution and reloading (P. Sosnitza, M. Farooqui, M. Saleemuddin, R. Ulber and T. Scheper 1988. Anal. Chim. Acta 368:197).

5 References

1. Khmelnitsky YL, Levashov AV, Klyachko NL, Martinek K (1988) Enzyme Microb Technol 10:710
2. Gupta MN (1992) Eur J Biochem 203:25
3. Gianfreda L, Scarfi MR (1991) Mol Cell Biochem 100:97
4. Gemeiner P, Rexova-Benkova L, Svec F, Norrlow O (1994) Natural and synthetic carriers suitable for immobilization of viable cells, active organalles and molecules. In: Veliky IA, McLean RJC (eds) Immobilized biosystems. Theory and practical applications. Blackie Academic and Professional Chapman and Hall, London p 1
5. Mattiasson B (1988) Methods Enzymol 137:647
6. Gupta, MN, Mattiasson B (1992) Unique applications of immobilized proteins in bioanalytical systems. In: Suelter CH (ed) Bioanalytical applications of enzymes. Wiley 36:1
7. Parker CW (1976) Radioimmunoassay of Biologically active compounds. Prentice-Hall Englewood Clifts, New Jersey
8. Shoemaker H, Wall M, Zurawski V (1984) Biotech 84. Online Pinner England, p 405
9. Hubbard AL, Cohen ZA (1976) In: Muddy AH (ed) Biochemical analysis of membrane. Wiley New York, p 427
10. Green NM (1963) Biochem J 89:585
11. Lamet D, Isenman D, Sjodahl J, Sjoquist J, Pecht I (1978) Biochem Biophys Res Commun 85:608
12. Vijayalakshmi MA (1989) Trends Biotechnol 7:71
13. deAlwis WU, Hill BS, Meiklejohn BI, Wilson GS (1987) Anal Chem 59:2688
14. Melchers F, Messer W (1970) Biochem Biophys Res Commun 40:570
15. Sato PH, Waltton DM (1983) Arch Biochem Biophys 221:548
16. Ikura K, Okumura K, Yoshikawa M, Sasaki R, Chiba H (1984) J Appl Biochem 6:222
17. Solomon B, Moav N, Pines G, Katchalski-Katzir E (1984) Mol Immunol 21:1
18. Solomon B, Koppel R, Pines G, Katchalski-Katzir E (1986) Biotechnol Bioeng 28:1213
19. Solomon B, Hollander Z, Koppel R, Katchalski-Katzir E (1987) Methods Enzymol 135:160
20. de Alwis U, Wilson GS (1989) Talanta 112:249
21. Fusek M, Turkova J, Stovickova J, Franek F (1988) Biotechnol Lett 10:85
22. Shami EY, Rothstein A, Ramjeesingh M (1989) Trends Biotechnol 7:186
23. Ruoff P, Lillo C, Campbell WH (1989) Biochem Biophys Res Commun 161:496
24. Shami EY, Ramjeesingh M, Rothstein A, Zywulko M (1991) Enzyme Microb Technol 13:424
25. Solomon B, Koppel R, Schwartz F, Flemminger G (1990) J. Chromatogr 510:321
26. Stovickova J, Franek F, Turkova J (1991) Biocatalysis 5:121
27. Agnellini D, Pace M, Cenquanta S, Gardana C, Pietta PG, Mauri PL (1992) Biocatalysis 6:251
28. Jafri F, Husain S, Saleemuddin M (1993) Biotechnol Appl Biochem 18:401
29. Jafri F, Husain S, Saleemuddin M (1995) Biotechnol Tech 9:117
30. Jafri F, Saleemuddin M (1997) Biotechnol Bioeng 56:605
31. Siemann M, Syldatk C, Wagner F (1994) Biotechnol Lett 16:349
32. Suzuki T, Pelichova H, Cindader B (1996) J Immunol 103:1366.
33. Sada E, Katon S, Kiyokawa A, Kondo A (1988) Biotechnol Bioeng 31:635
34. Arnon H, (1973) Immunochemistry of enzymes. In: Sela M (ed) The antigens vol I. Academic Press, New York, p 88
35. Cinader B (1967) Antibodies to biologically active molecules. Pergamon Press, Oxford, p 88
36. Zyk N, Citri N (1968) Biochim Biophys Acta 139:317
37. Michaeli JD, Pinto E, Benjamini, deBuren FP (1969) Immunochemistry 6:101
38. Feinstein RN, Jaroslow BN, Howard JB, Faulhaber JT (1971) J Immunol 106:1316
39. Ben-Yosef Y, Geiger B, Arnon R (1975) Immunochemistry 12:221
40. Lis H, Sharon N (1993) Eur J Biochem 218:1
41. Feizi T, Childs RA (1990) Biochem J 245:1

42. Zopf DA, Tsai CM, Ginsburg V (1978) Arch Biochem Biophys 185:61
43. Hansen RS, Beavo JA (1982) Proc Natl Acad Sci USA 79:2788
44. Ernst-Cabrera K, Wilchek M (1988) Trends Anal Chem 7:58
45. Guisan JM, Bastida A, Cuesta C, Fernandez-Lawfente R, Russel CM (1991) Biotechnol Bioeng 38:1144
46. Iqbal J, Saleemuddin M (1983) Biotechnol Bioeng 13:641
47. Solomon B, Balas N (1991) Biotechnol Appl Biochem 14:202
48. Goding JM (1985) Monoclonal antibodies: principles and practices. Academic Press, New York, p 9
49. Capra JD, Edmundson AB (1977) Scientific Am 236:50
50. Vokley J, Harris J (1984) Biochem J 217:535
51. Folkersen J, Tiesner B, Westergaard JG, Grudzinskas JG (1985) J Immun Methods 77:45
52. Desai MA (1990) J Chem Tech Biotechnol 48:105
53. Jack GW, Black R, James K, Boyel JE, Micklein LR (1987) J Chem Tech Biotechnol 39:45
54. Bezin H, Melache JM (1986) J Immunol Methods 88:19
55. Tan-Wilson AL, Reichin M, Noble RW (1978) Immunochemistry 13:921
56. Stankus RP, Leslie GA (1976) J Immunol Methods 10:307
57. Kristiansen Y (1978) In: Hoffmann-Ostenhof, O (ed) Matrix bound antigens and antibodies in affinity chromatography. Pergamon Press, p 191
58. Kohno H, Kanda S, Kanno T (1986) J Biol Chem 261:10744
59. Emomoto A, Kamata N, Nakamura K (1994) Biochem Biophys Res Commun 201:1008
60. Husain Q, Iqbal J, Saleemuddin M (1985) Biotechnol Bioeng 27:1102
61. Kokufuta E, Yamaya Y, Shimada A, Nakamura I (1988) Biotechnol Lett 10:301
62. Cuatrecasas P, Wilchek M, Ainfinsen CB (1968) Proc Natl Acad Sci USA 61:636
63. Campbell DL, Weliky N (1967) In: Willams CA, Chase MW (eds) Methods in Immunology and Immunochemistry. Academic Press, New York, p 365
64. Silman TH, Katchalski E (1986) Ann Rev Biochem 35:873
65. Ehle H, Horn A (1990) Biosepration 1:97
66. Kondo A, Yamaski R, Higashitani K (1992) J Ferment Bioeng 74:226
67. Kondo A, Kaneko T, Higashitani K (1994) Biotechnol Bioeng 44:1
68. Wang P, Hill TG, Wartchnow CA, Hnston ME, Oechler LM, Smith MB, Bednarski MD, Callstrom MR (1992) J Am Chem Soc 114:378
69. Youings A, Chang SC, Dwek RA, Scragg IG (1996) J Biol Chem 314:621
70. Quash G, Roch AM, Nivelean, Grange J, Kevlouangkhot T, Huppert J (1978) J Immunol Meth 22:165
71. Prisyazhnoy VS, Fusek M and Alkhov B (1988) J Chromatogr 424:243
72. Hoffman WL, O'Shannessy DJ (1988) J Immunol Meth 112:113
73. Little MC, Siebert CJ, Matson RS (1988) Biochromatogr 3:156
74. Fleminger G, Hadas E, Wolf T, Soloman B (1990) Appl Biochem Biotechnol 23:123
75. Brizgys MV, Pincus SH, Rollins DE (1988) Biotechnol Appl Biochem 10:373
76. Zemek J, Kuniak P, Gemeiner J, Zamocky J, Kucar S (1982) Enzyme Microb Technol 4:233
77. Wormald MR, Rudd PM, Harvey DJ, Chang SC, Scragg IG, Dwek RA (1997) Biochemistry 36:1370
78. Schneider C, Newman RA, Sulkerland DR, Asser V, Greaves MF (1982) J Biol Chem 257:10766
79. Solomon B, Koppel R, Katchalski-Katzir E (1984) Bio/Technology August:709
80. Hale JE, Beidler DE (1994) Anal Biochem 222:29
81. Hale JE (1995) Anal Biochem 231:46
82. deAlwis U, Wilson GS (1987) Anal Chem 59:2786
83. deAlwis WU, Wilson GS (1985) Anal Chem 57:2754
84. Janatova J, Gobel RJ (1984) Biochem J 221:113
85. Kondo A, Kishimura M, Katoh S, Sada E (1989) Biotechnol Bioeng 34:532
86. Scheriff S, Silverton EW, Padlan EA, Cohen GH, Smith-Gill SJ, Finzel BC, Davies DR (1987) Proc Natl Acad Sci USA 84:8075
87. Jones S, Thornton JM (1996) Proc Natl Acad Sci USA 93:13

88. Davies DR, Sheriff S, Pardon EA (1988) J Biol Chem 263:10541
89. Rees AR, Roberts S, Webster D, Cheetam JC (1988) In: Brew K, Ahmad F, Baily H et al. (eds) ICSU Short reports V8, IRL Press, p 172
90. Tanford C (1990) Adv Protein Chem 24:1
91. Sadana A, Madgula A (1993) Biotechnol Prog 9:259
92. Solomon B, Schwartz F (1995) J Mol Recog 8:72
93. Carlson JD, Yarmush ML (1992) Bio/Technology 86
94. Katzav-Gozansky T, Hanan E, Solomon B (1996) Biotechnol Appl Biochem 23:227
95. Daniel RM, Cowan DR, Morgan HW, Curran MP (1982) Biochem J 207:641
96. Reslow M, Adlercreutz P, Mattiasson B (1988) Eur J Biochem 172:573
97. Whetze E, Adlercreutz P, Mattiasson B (1992) In: Velmine TJ, Beeflink MH, Von-Stokar V (eds) Biocalalysis in non-conventional media (Progress in Biotechnology V 8), p 372
98. Russel AJ, Trudel LJ, Skipper PL, Groopman JD, Tannenbaum SR, Klibanov AM (1989) Biochem Biophys Res Commun 158:80
99. Wang P, Hill TG, Wartchow CA, Huston ME, Oehler LM, Smith MB, Bednarski MD, Callstorm MR (1992) J Am Chem Soc 114:378
100. Janda KD, Ashley JA, Jones TM, MeLeod DA, Schloeder DM, Weinhouse MI (1990) J Am Chem Soc 112:8886
101. Arbige MV, Pitcher WH (1989) Trends Biotechnol 7:330
102. Sharon N (1993) Trends Biochem Sci 18:221
103. Saleemuddin M, Husain Q (1991) Enzyme Microb Technol 13:290
104. Goldstein IJ (1988) Studies on the combining sites of Concanavalin A. In: Chaudhury TK, Wiess AK (eds) Concanavalin A Plenum Press, New York, London, p 35
105. Goldstein IJ, Hollermann CE, Merrick JM (1965) Biochim Biophys Acta 97:68
106. Kobata A (1992) Eur J Biochem 209:483
107. Ogata S, Muramatsu I, Kobata A (1975) J Biochem (Tokyo) 78:687.
108. Reeke GN, Jr, Becker JW, Cunninaham BA, Wang JL, Yahara I, Edelman GM (1975) Structure and function of Concanavalin A. In: Choudhury TK, Weiss AK (eds) Concanavalin A. Plenum Press, New York, London, p 43
109. Mandal DK, Brewer CF (1993) Biochemistry 32:5116
110. Ahmad A, Bishayee S, Bacchawat BK (1973) Biochem Biophys Res Commun 53:730
111. Husain Q, Saleemuddin M (1986) Enzyme Microb Technol 8:686
112. Gunther GR, Wang JL, Yahara I, Cunniggham BA, Edelman GM (1973) Proc Natl Acad Sci USA 70:1012
113. Fadda MB, Rescigno A, Rinaldi A, Sanjust E (1992) Biotechnol Appl Biochem 16:221
114. El Rassi Z, Truei Y, Maa Y-H, Horvath C (1988) Anal Biochem 169:1722
115. Anspach FB, Altmann-Haase G (1994) Biotechnol Appl Biochem 20:323
116. Farooqi M, Saleemuddin M, Ulber R, Sosnitza P, Scheper T (1997) J Biotech 55:171
117. Gemeiner P, Docolomansky P, Nahalka J, Stefuca V, Danielsson B (1996) Biotechnol Bioeng 49:26
118. Vrabel P, Polakovic M, Godo S, Bales V, Docolomansky P, Gemeiner P (1997) Enzyme Microb Technol 21:196
119. Montero MA, Remeu A (1993) Biochem Mol Biol Int 30:685
120. Hsiao H, Royer GP (1978) Arch Biochem Biophys 198:379
121. Woodward J, Wiseman A (1978) Biochem Biophhys Acta 527:8
122. Ulbrich-Hofman R, Golbik R, Damerau W (1993) Fixation of the unfolding region -a hypothesis of enzyme stabilization. In: Vander Tweel WJJ, Harder A, Buitelaar RM (eds) Stability and stabilization of enzymes. Elsevier, p 497
123. Arnold U, Rucknagel KP, Shierhorn A, Ulbrich-Hofman R (1996) Eur J Biochem 237:862
124. Goldstein IJ (1976) Carbohydrate binding specificity of concanavalin A. In: Bittiger H, Schnebli HP (eds) Concanavalin A as a tool. Wiley, New York, London, p 55
125. Mislovicova D, Vikartovska A, Gemeiner P (1997) J Biochem Biophys Methods 35:37
126. Surolia A, Bishayee S, Ahmad A, Balasubramanian KA, Thambi-Dorai D, Poddar SK, Bacchawat BK (1975) Studies on the interaction of concanavalin A with glycoproteins. In: Chowdhuri TK, Weiss AK (eds) Concanavalin A. Plenum Press, New York, London, p 95

127. Mislovicova D, Chudinova M, Gemeiner P, Docolomansky P (1995) J Chromatogr B 664:145
128. Mattiasson B, Borrebaeck C (1975) FEBS Lett 85:119
129. Mattiasson B, Danielsson B (1982) Carbohydr Res 102:273
130. Koneke R, Menzel C, Ulber R, Schiigerl K, Scheper T, Saleemuddin M (1996) Biosensors & Bioelectronics 11:1229
131. Mattiasson B, Johansson PA (1982) J Immunol Methods 52:233
132. Mattiasson B (1983) In: Immobilized cells and organalles. CRC Press Boca Raton, Florida, p 95
133. Kaul R, D'Souza SF, Nadkarni GB (1986) J Microb Technol 1:12
134. Habibi-Rezaei M, Nemat-Gorgani M (1997) Appl Biochem Biotechnol 67:99
135. Siddiqui S, Hasan S, Salahuddin A (1995) Arch Biochem Biophys 419:426
136. Siddiqui S, Anwar A, Saleemuddin M. unpublished observations
137. Porath J, Olin B (1983) Biochemistry 22:1621
138 Porath J (1988) Trends Anal Chem 7:254
139. Porath J, Carlsson J, Olsson I, Belfrage G (1975) Nature (London) 258:598
140. Arnold FH (1991) Bio/technology 9:151
141. Sulkowski E (1989) Bio Essays 10:169
142. Sousa C, Cebolla A, deLorenzo V (1996) Nature Biotechnology 14:1017
143. Hemdan ES, Zhao Y-J, Sulkowski E, Porath J (1989) Proc Natl Acad Sci USA 86:1811
144. Sundberg RJ, Martin RB (1974) Chem Rev 74:471
145. Yip TT, Nakagawa Y, Porath J (1989) Anal Biochem 183:159
146. Klapper MH (1977) Biochem Biophys Res Commun 78:1018
147. Sulkowski E (1985) Trends Biotechnol 3:1
148. Wong JM, Albright RL, Wang NHL (1991) Sep Purif Meth 20:49
149. Ljungquist C, Breitholtz A, Brink-Nilsson H, Moks T, Uhlen M, Nilsson B (1989) Eur J Biochem 186:563
150. Piesecki S, Teng W-Y, Hochuli E (1993) Biotechnol Bioeng 42:178
151. Hochuli E, Dobeli H, Schacher A (1987) J Chromatogr 411:177
152. Carlsson J, Mosbach K, Bulow L (1996) Biotechnol Bioeng 51:221
153. Chaga C (1994) Biotechnol Appl Biochem 20:43
154. Coulet PR, Carlsson J, Porath J (1981) Biotechnol Bioeng 23:663
155. Anspach FB, Altmann-Hasse G (1994) Biotechnol Appl Biochem 20:313
156. McKenzie GH, Sawer WH, Nichol LW (1972) Biochim Biophys Acta 263:283
157. Turkova J (1993) In: Bioaffinity Chromatography, J Chromatography Library, vol 55. Elsevier 55:644
158. Sholtissek S, Grosse F (1988) Gene 62:55
159. diGuan C, Li P, Riggs PD, Inouye H (1988) Gene 67:21
160. Langsford ML, Gilkes NR, Wakarchuk WW, Kilburn DG, Miller RC, Jr, Warren RAJ (1984) J Gen Microbiol 130:1367
161. Gilkes NR, Warren RAJ, Miller RS, Jr, Kilburn DG (1988) J Biol Chem 263:1040
162. Ong E, Gilkes NR, Warren RAJ, Miller RC, Jr, Kilburn DJ (1989) Bio/Technology 7:604
163. Ong E, Gilkes NR, Miller RC, Jr, Warren RAJ, Kilburn DC (1991) Enzyme Microb Technol 13:59
164. Phelps MR, Hobbs JB, Kilburn DJ, Turner RFB (1994) Biotechnol Progr 10:433
165. Baneyx F, Schmidt C, Georgiou G (1990) Enzyme Microb Technol 12:337
166. Newsted WJ, Ramjeesingh M, Zywulko M, Rothstein SJ, Shami EY (1995) Enzyme Microb Technol 17:757

Author Index Volume 1–64

Author Index Vols. 1–50 see Vol. 50

Subject Index

GPSR Compliance

The European Union's (EU) General Product Safety Regulation (GPSR) is a set of rules that requires consumer products to be safe and our obligations to ensure this.

If you have any concerns about our products, you can contact us on ProductSafety@springernature.com

In case Publisher is established outside the EU, the EU authorized representative is:

Springer Nature Customer Service Center GmbH
Europaplatz 3
69115 Heidelberg, Germany

MIX
Papier | Fördert gute Waldnutzung
FSC® C083411

Kolibri 360 GmbH
Ferdinand-Jühlke-Straße 7
99095 Erfurt, Deutschland
produktsicherheit@kolibri360.de